THE CARBON TAX QUESTION

THE CARBON TAX QUESTION

CLARIFYING CANADA'S MOST CONSEQUENTIAL POLICY DEBATE

THOMAS F. PEDERSEN

HARBOUR PUBLISHING

COPYRIGHT © 2024 THOMAS F. PEDERSEN

1 2 3 4 5 — 28 27 26 25 24

ALL RIGHTS RESERVED. No part of this publication may be reproduced, stored in a retrieval system or transmitted, in any form or by any means, without prior permission of the publisher or, in the case of photocopying or other reprographic copying, a licence from Access Copyright, www.accesscopyright.ca, 1-800-893-5777, info@accesscopyright.ca.

HARBOUR PUBLISHING CO. LTD.
P.O. Box 219, Madeira Park, BC, VON 2H0
www.harbourpublishing.com

EDITED by Michael Barclay
INDEXED by François Trahan
COVER DESIGN by John Montgomery
TEXT DESIGN by Libris Simas Ferraz | Onça Publishing
PRINTED AND BOUND in Canada
PRINTED on 100% recycled, FSC®-certified paper

HARBOUR PUBLISHING acknowledges the support of the Canada Council for the Arts, the Government of Canada, and the Province of British Columbia through the BC Arts Council.

CATALOGUING DATA AVAILABLE FROM LIBRARY AND ARCHIVES CANADA
Title: The carbon tax question : clarifying Canada's most consequential policy debate / Thomas F. Pedersen.
Names: Pedersen, Thomas F., 1951- author
  https://id.oclc.org/worldcat/entity/E39PCjqqgYfWhVjd9JWyvxYQMP
Description: Includes bibliographical references and index.
Identifiers: Canadiana (print) 2024041506X | Canadiana (ebook) 20240415094 | ISBN 9781990776977 (softcover) | ISBN 9781990776984 (EPUB)
Subjects: LCSH: Carbon taxes—British Columbia—History. | LCSH: Carbon taxes—Political aspects—British Columbia. | LCSH: Carbon taxes—Economic aspects—British Columbia.
Classification: LCC HJ5376.B7 P43 2024 | DDC 336.2/78333709711—dc23

*For Jenny, Charlotte and David.
I am proud of how you are
helping to shape
a better future.*

***

*And to Jimmy:
May that future offer
you and all other grandchildren
climate stability and endless
opportunity.*

Table of Contents

PROLOGUE

ix

CHAPTER 1

How the Mountain Pine Beetle Brought Global Warming to the Forefront in British Columbia

1

CHAPTER 2

The Rise of Climate Action

The Role of Gordon Campbell

14

CHAPTER 3

British Columbia's Carbon Tax

"A template for the world"

37

CHAPTER 4

Down Under I:
The Hawke-Keating-Howard Years and Australian Climate Action Dithering

66

CHAPTER 5

Down Under II:
Yo-yo Politics and Australian Carbon Pricing

The Rudd-Gillard Years

86

CHAPTER 6

Down Under III: Yo-yo Politics and
Australian Carbon Pricing

The Demise

111

CHAPTER 7

Cutting Off One's Nose to Spite One's Electoral Prospects

The 2008–09 Campaign to "Axe the Tax" in British Columbia

134

CHAPTER 8

Going Wobbly on Climate Action

The Christy Clark Years in British Columbia, 2011–17

149

CHAPTER 9

How to Sell a Carbon Tax

Three Stories

186

CHAPTER 10

And Now to Canada:
Have We Not Yet Learned?

195

Endnotes · 211

Acknowledgements · 236

Index · 239

About the Author · 252

Prologue

BRIAN MENOUNOS KNOWS ICE. THE HOLDER OF A CANADA Research Chair at the University of Northern British Columbia, Menounos has spent over two decades studying the behaviour of BC glaciers in the face of a changing climate.

It's disquieting research. Using archived aerial photographs and satellite images, he and his team showed that the more than fifteen thousand glaciers in the province lost some 11 per cent of their area in just the twenty years between 1985 and 2005.[1] At this rate, the calendar photographs of majestic glaciers that so many British Columbians grew up admiring will be mere historical artifacts by the end of this century. Glacier National Park, in the southeast of the province, will be a misnomer.

We know why this happening: Earth is warming as a direct consequence of adding to the atmosphere greenhouse gases—carbon dioxide and others—derived from the burning of fossil fuels, deforestation and agriculture. The impacts of rising temperatures are vivid, almost universally negative—and everywhere. Shame on us, for we were warned almost one hundred and sixty years ago that this would likely happen.

In the late 1850s, John Tyndall, professor of natural philosophy at the Royal Institution of Great Britain, hypothesized that certain atmospheric gases absorbed heat attempting to escape from the sunlight-warmed surface of the Earth. That notion had been advanced some three decades earlier by the famous French mathematician Jean-Baptiste Joseph Fourier, but had never been tested. In tackling his hypothesis, Tyndall built an apparatus that measured absorption of infrared radiation by gases contained in an enclosed cylinder. He showed, for the first time, that carbon dioxide, water vapour and methane (among others) were effective at absorbing—and re-emitting—"radiant heat," or infrared radiation. There was self-interest in his focus on this topic: Tyndall was a keen mountaineer who wanted to understand glacial behaviour in the Alps, and its relationship to climate.

In 1859 he presented his results on the absorption of radiant heat in an evening lecture at the Royal Institution, noting:

> *The bearing of this experiment upon the action of planetary atmospheres is obvious ... the atmosphere admits of the entrance of the solar heat, but checks its exit; and the result is a tendency to accumulate heat at the surface of the planet.*[2]

"Checks its exit." Those three words, delivered over a century and a half ago, capture the first experimentally verified explanation of the greenhouse effect. In a paper just two years later, Tyndall foreshadowed future global warming when he pointed directly at atmospheric carbon dioxide and methane as agents for warming the planet.

It was 1861. We were warned.

The warnings intensified in the 1970s. The phrase *global warming* entered the modern lexicon via Wallace Broecker, professor at Columbia University in New York and a giant in the field of ocean and climate science. In 1975, he posed this question in the leading journal *Science*: "Climatic Change: Are We on the Brink of a Pronounced Global Warming?"[3]

The answer is now more than obvious: yes. Since the 1970s, global average temperature has continued to rise in close association with ever-increasing concentrations of greenhouse gases in the atmosphere—particularly carbon dioxide, but also methane and nitrous oxide.

The effects of higher temperatures are now so obvious as to bring justified scorn on the shrinking number of deniers who troll the internet or, occasionally, stalk the halls of Congress: rising sea levels; disappearing Arctic summer sea ice; the satellite-measured loss of some three hundred billion tonnes of ice per year from the Greenland ice cap; increased frequency and intensity of rainfall deluges and heat waves; prolonged regional droughts on scales not seen in the last thousand or more years; poleward migration of plant and animal species; the progressive loss of permafrost in northern latitudes; and disappearing alpine glaciers in every corner of the world.

Some jurisdictions have moved to blunt the effects of global warming, either by attempting to reduce greenhouse gas emissions, or by adapting, or both. One of them was British Columbia. Why did this province act when others procrastinated? What drove British Columbians to adopt North America's first broad-spectrum carbon tax in 2008? Why BC, of all places?

British Columbia was mostly successful in at least beginning to address John Tyndall's challenge. But on the other side of the planet, a parallel effort largely failed. Australians had high hopes after their 2007 national election that they could slow the rate at which they were belching greenhouse gases into the atmosphere. And for a brief while they did make progress. But a political revolving door saw well-meaning leaders replaced by at least one irresponsible leader, who denied science and was a darling of the tabloid media. And no, his surname was not Trump.

The remarkable steps that BC undertook in the late 2000s reflect a rare convergence of external and internal influences, combined with thoughtful leadership, smart politics and vision. It's a story of ups and downs, of ideological inconsistencies, of adhering to principles or subverting them. And while British Columbia is the lead actor in the story, Australia and Canada's federal government play important supporting roles.

The story begins not with politics but with biology, for it was the actions of a little round-headed insect that got the attention of politicians two decades ago. The size of a matchhead, that insect wreaked economic, social and environmental havoc on forests in the British Columbia Interior, even as alpine glaciers were slowly melting.

Global warming was coming home to roost.

CHAPTER 1

How the Mountain Pine Beetle Brought Global Warming to the Forefront in British Columbia

It was a glorious big-sky afternoon in late July 2005, when Lester Johnson headed west from Edson, Alberta, hoping to make Prince George by dark. He was in a bit of a rush. As he navigated the eastern foothills of the Rockies, the *ping-ping-ping* echoing around the cab of his scruffy old v6 pickup was telling him not to push so hard on the gas pedal. Lester didn't care about the pings, because he couldn't hear them very well—the engine knock was muffled by a zillion whacks as clouds of six-legged black specks smacked into the windshield, spilling red-brown guts and gumming his wipers with slime. "It was like the sky had opened up and dumped an intense black rain on me. It was scary. I thought, 'What the hell *is* this?'"

Tales like his ricochet around forestry offices, taverns and homes in northern British Columbia to this day. What Lester was driving through became common in northern summers in the mid-2000s. The mountain pine beetle, or *Dendroctonus ponderosae*, an insect the size of a sesame seed or small grain of rice (Figure 1), was on the

CHAPTER 1

FIGURE 1. An adult mountain pine beetle (*Dendroctonus ponderosae*) perched on a matchstick. *Jeff Foott / Minden Pictures / Nature in Stock*

march. Living primarily in old and often weak or damaged pines, *D. ponderosae* was a relatively innocuous endemic species that ranged from northern British Columbia to northern Mexico. In previous centuries, populations would infrequently infest trees over discrete stands that might extend to thousands of hectares, particularly in south-central BC.[1] But that was no longer the case.

The scale of impact changed in the mid-1990s, when beetle populations exploded "almost simultaneously in lodgepole pine stands across the north central and southern part of the province."[2] As those infestations coalesced and grew over the following twenty years, BC's forested heartland changed from a dappled green tapestry to a rust- and grey-stained canvas where, too often, massive swaths of trees were dead or dying as far as the eye could see. Foresters, mill-workers, journalists, bureaucrats, farmers, hunters, fishermen, politicians and environmentalists—indeed, just about every citizen in British Columbia's Interior—took notice of the "lodgepole tsunami."[3] "What in the world is happening?" they asked.

Ken White, an entomologist and pine beetle expert for BC's provincial government, vividly recalls those early stages of the epidemic.[4] "It was late summer. We were flying in a Jet Ranger helicopter some two thousand metres above Tweedsmuir Park in west-central BC. We started hearing intermittent *ping-ping* sounds on the windscreen that became a constant patter, as thousands of specks smacked into the Plexiglas and coated it with a reddish-brown paste. I was trying to figure out what we were flying through, when I saw one of the specks smear across the rear bubble window. It was a mountain pine beetle. And then we flew out of the swarm and the patter suddenly stopped. It was unreal."

At about the same time, Peter Jackson was also witnessing migrating swarms of beetles, but he wasn't in a helicopter or even in a stand of pine trees. A professor of meteorology at the University of Northern British Columbia (UNBC), Jackson was monitoring precipitation radar imagery on his computer screen. One afternoon in July 2005, he was stunned by what he was seeing: a cloud on a clear-sky day, heading east from the Prince George region toward the Rockies. What the radar had detected was a massive swarm of adult mountain pine beetles, migrating by the billions in search of proverbial greener pastures—in this case, pine trees that weren't yet dead. Sometimes the wind would carry them over fields where they'd run out of steam without something to latch onto. Jackson's colleague at UNBC, forest entomologist Staffan Lindgren, remembers hearing stories of beetles hitting metal roofs in the Peace River region, "sounding like rain, but on perfectly clear days."[5]

It was the numbers of beetles that truly impressed then. In one instance, billions of carcasses washed ashore at Frank Lake near Vanderhoof. As Ken White describes the scene, the masses of beetles were inches thick, draped over logs and masses of lodgepole pine needles that ran down the shoreline. "It appears they were trying to fly over the lake, and, for whatever reason, didn't make it across." Local trout were happy—they gorged on the drowning hordes. White says that when the chief forester, Larry Pedersen, was taken on a tour in the Tweedsmuir area, "we landed at a camp for lunch, and the crew had caught fresh trout for us. They said the guts were completely packed with mountain pine beetles."

Those billions—or more accurately, trillions—of beetles wreaked havoc across the province. By 2016, lodgepole and ponderosa pines spanning more than nineteen million hectares—an area four times the size of Nova Scotia—had been ravaged in BC's Interior forests, a staggering, unprecedented tally (Figure 2). It's a tally that forever changed the climate-action political landscape in BC.

The tale of why this happened weaves biology, forest management, fire suppression and climate change together into a story of massive economic and ecological dislocation and destruction.

That story begins with biology.

Mountain pine beetles spend most of their one- to two-year lifespans hidden from view; the first three of their four life stages—egg, larva, pupa—are spent in the fibrous, soft inner bark (a.k.a. the phloem layer) under the hard outer bark of host trees. In late spring to early summer, the pupae mature into adults, bore their way through the outer bark, and in July or August fly off to infest new trees.

As fliers, the beetles are not Spitfires. They run out of steam fairly quickly, and only manage to hit a top speed akin to a brisk walk. Despite those limitations, they can sometimes manage to jump from one pine stand to another, twenty or more kilometres away, by riding the wind or local convection currents. Indeed, it's been reported that they are sometimes carried as far as three hundred kilometres by strong winds before landing on their next victims.[6] The speed at

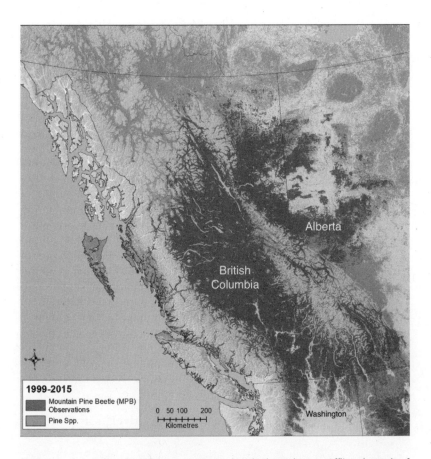

FIGURE 2. Area infested by the mountain pine beetle through 2015. Afflicted stands of trees are shown in the darkest grey. As of 2017, established infestations in north-central Alberta had pushed eastward to within about 215 kilometres of the Saskatchewan border. *Natural Resources Canada, Canadian Forest Service*

which they spread across British Columbia summer after summer in the 2000s indicates that they took full advantage of such tailwinds.

Once airborne, typically when daytime highs exceed 16° C, adult beetles seek out new hosts using a combination of smell—at which they are really good—and vision, at which by all accounts they are as deficient as Mr. Magoo. Their rudimentary eyes apparently allow them to make out wide, vertical silhouettes in forests; those become the preferred landing targets. Once an adult beetle has grabbed onto the trunk of a tree, it senses odours emitted by the bark, and decides whether to stay or move on to another silhouette offering a more appropriate perfume.

That's when the attack on a mature pine becomes really interesting. Once a female has settled on a tree she deems suitable, and starts to bore into the bark, she emits pheromones—chemical signals—that waft through the air. That's an invitation to hordes of other beetles, both male and female, which then join her in mounting a mass attack. This can last for a day or two, and up to sixty beetles per square metre of bark area can be involved, all beavering away, drilling through the hard outer bark to reach the underlying soft phloem a few millimetres away.

That up-to-sixty limit is important: the beetles have evolved to understand that their pine buffet can support only a finite density of diners, and sixty per square metre it is. To keep control of the numbers, it is thought that both males and females release different pheromones—"termination" compounds—that redirect incoming beetles to nearby trees.

Mature pine trees are stubborn, though. They fight back in the face of an attack, releasing resins from ducts severed by the beetles (Figure 3). That viscous, sticky resin can slow the actions of the invading beetles, and may even repel them if it saturates the phloem quickly. As it seeps out of the freshly bored entrance holes, the resin congeals,

FIGURE 3. Popcorn-like masses of pitch on pine bark, the result of the tree attempting to repel beetles by flooding entrance holes with pitch. USDA *Forest Service* / *Wikimedia Commons* / *Public Domain*

forming distinctive, gnarly, popcorn-like pinkish-white blobs of pitch on the bark's exterior.

Mass attacks are the most effective tool that beetles use to overwhelm these defensive efforts. When a female manages to dodge the flood of pitch and makes it through the outer bark, she uses her strong jaws to chew a brooding gallery in the phloem, parallel to the trunk, that can be as much as ninety centimetres long. Once she mates, she lays an average of five dozen millimetre-sized eggs in little niches cut into the sides of the gallery. The broods hatch one to two weeks later, and the larvae commence chewing channels sideways. The resulting primary galleries with side channels are rather elegant, looking like the fronds of young ferns (Figure 4), but the effect on the trees is devastating. The feeding patterns of the larvae effectively girdle the trunk, preventing vertical fluid transfer. It is that circulatory collapse that eventually leads to the death of the tree—typically in one or two years.

Larvae overwinter under the bark. They slow or even stop feeding when temperatures fall as winter approaches. Survival through the season is a bit of a lottery—or at least it used to be. In a striking evolutionary adaptation, pinebeetle larvae begin producing an antifreeze, in this case glycerol, as autumn arrives and temperatures decline. The glycerol accumulates in their blood. If they have sufficient time over October and November, they produce enough of it to survive mid-winter temperatures as low as -29 or even down to -40° C.

That's a piece of biological magic. But should cold snaps arrive in the BC Interior around the end of October, or again in mid-March after the larvae have begun to metabolize the glycerol, temperatures below about -23° C can be lethal.

Global warming has disrupted survival and mortality patterns that have been in place for many centuries. Nowadays, British Columbia's Interior plateau rarely sees the legendary autumn cold snaps of the past. Prior to about the mid-1990s, minimum October

FIGURE 4. Mountain pine beetle damage beneath the bark of a lodgepole pine tree in western Alberta. Vertical troughs are excavated by adult females and horizontal channels are bored by larvae. *Katherine Bleiker, Natural Resources Canada*

or early November temperatures in central BC often plunged to the minus-twenties. In 1984–85, for example, an outbreak of the mountain pine beetle in the Chilcotin region was stopped in its tracks by -30° C lows in late October and early November.[7] But mid-autumn temperature minima over the last thirty years from Interior sites, registered in the Provincial Climate Data Set, rarely show such values.[8] Put simply, British Columbia was getting warmer on either side of the new millennium. Beetle survival soared.

Larvae that survive the winter enlarge their feeding tunnels in the spring, and create chambers in which they graduate to the pupal stage. A few weeks later, they metamorphose into young beetles that feed not just on phloem, but also on the spores of a blue-stain fungus. That fungus gets into the trees by stowing away in little pockets on the heads of invading adults. Mountain pine beetles have a mutually beneficial relationship with that fungus: the spores multiply and coat the interior walls of pupal chambers, providing a rich food source that nourishes the young beetles prior to their emergence as mature adults. In turn, the fungus gets to colonize the sapwood of the tree, imparting a distinctive grey-blue mottled texture to the wood.

The blue hue was first thought by lumber marketers to be a commercial disaster. They loathed the idea of trying to sell "bug-killed wood" tainted with an unusual blue tinge. The power of marketing soon changed that. In a moment of entrepreneurial ingenuity in 2001, just when mill owners were throwing up their hands in despair at the prospect of trying to sell masses of blue-tinged pine, Lynn Pont in Quesnel, BC, had an inspiration.[9] Instead of calling it beetle-kill wood, she thought, why not label it "Denim Pine" and make it available to furniture and flooring companies and others as a highly novel product? She trademarked the name, set up a marketing association, and made history. Denim Pine became both widely available and prized for its unique textural beauty.

But while growing interest in blue-mottled pine may have given some an impression of the epidemic's silver lining, that impression was short-lived—it paled in comparison with the scale of the disaster that the epidemic imposed on the Interior. No doubt global warming helped drive that epidemic to epic proportions.

And it has not yet run its course. *Dendroctonus ponderosae* jumped both the Rockies *and* its preferred tree species several years ago and is now chewing its way through the boreal jack-pine forests of northern Alberta. Its range also shifted northward. In 2011 the beetle was discovered within eighty kilometres of the Yukon border,[10] and in 2012 was observed for the first time in the southwestern corner of the Northwest Territories.[11] Some authorities suggest that the beetle will flit from stand to stand eastward across the vast northern boreal forest over the course of the coming decades—how fast that might happen and how far it might go remain open questions.[12]

One other factor has been at play here: wildfire suppression, which has improved at limiting the spread of fire to preserve the commercial stands of timber, and to prevent Interior communities from going up in smoke.

Those suppression efforts paid off, but at a cost draped in irony. Fighting forest fires saved harvestable pine trees and increased the stock of commercially valuable mature trees. Indeed, it has been estimated that the area of mature pine increased threefold during the twentieth century.[13]

Those trees were perfect candidates for invasion by *D. ponderosae*. Our past unwillingness to allow nature to take its course—by letting fires rejuvenate Interior forests—inadvertently created a landscape that, to the beetle, looked just like an "all you can eat" buffet. And when climate change is thrown into the mix—with its drier, hotter summers and shorter, warmer winters—managing the now-mature forest becomes a task of Herculean proportions.

There is now no doubt that the scale of the epidemic was exacerbated by putting a standing stock of large-diameter pines—unexposed to fire for several decades—in front of an exploding pine-bark beetle population.

The impact was profound. Between 1999 and 2017, the beetles killed some seven hundred and fifty million cubic metres of pinewood—60 per cent of the provincial standing stock of lodgepole and ponderosa pine—with a value in excess of $100 billion. The government responded by dramatically increasing the allowable annual cut, and industry raced to harvest dead-but-still-standing trees before decomposition set in.

For a brief few years, it was a gold rush. The processing capacity of some mills was enhanced, not by hiring more workers but rather by mechanizing. But as the wood supply began to collapse, some sawmills closed, unemployment rose, and the heartland of British Columbia—where nearly half of the regional economy depends on harvesting trees and milling lumber or producing pulp—began to hurt.

In 1990, 88 mills processed 42 million cubic metres of wood in the Interior of British Columbia. In the peak year of 2006, while the pine gold rush was in full swing, 72 remaining mills sawed 52 million cubic metres of wood, but by 2014 only 54 mills were still in operation, and in that year only 39 million cubic metres were cut into lumber.[14] Interior communities were hit hard by job losses as the regional fibre supply evaporated. Merritt, for example, a town of eight thousand residents some 270 kilometres northeast of Vancouver, saw four of its five sawmills close. The last of those shutdowns was announced in September 2016, following "the end of elevated cut levels for pine beetle-killed lumber."[15] That closure threw two hundred employees out of work. In a stroke of timing that dispirited entire families, the pink slips were handed out just before Christmas.

The socio-economic effects will be long-lasting. In 2016, the BC Ministry of Forests, Lands and Natural Resource Operations concluded that Interior "harvest levels are expected to decrease to about 20 per cent below pre-infestation levels within ten years, and this reduction could last five decades."[16] The costs to BC's treasury over the course of the epidemic are difficult to calculate, but likely run into the billions of dollars. Lost stumpage revenues were immense;[17] in 2000 the stumpage fee in the Prince George Forest Region was $31.28 per cubic metre of harvested logs.[18] But to encourage rapid harvesting, the fee for beetle-infested pinewood in subsequent years was reduced to a mere twenty-five *cents* per cubic metre. Given an annual allowable cut that was accelerated by tens of millions of cubic metres per year during the epidemic, forgone stumpage revenue ran to hundreds of millions of dollars annually. Adding to the fiscal penalty were the impacts of persistently high unemployment in the region, reduced consumer confidence and lower demand for goods and services.

All of that can be laid at the six tiny feet of an insect the size of a sesame seed, whose population exploded because human beings are warming up planet Earth.

The mountain pine beetle epidemic was Mother Nature's way of screaming that humans can't keep on burning fossil fuels as in the past. And while some remained deaf to her screams, few could ignore the hues of burnt sienna, rust and grey that so dramatically splashed across the heart of the BC landscape in the earliest years of this century.

In a fortunate and somewhat surprising turn of events, that visual wake-up call was heeded by the premier of the day, a man who listened, consulted and understood. And then, in a two-year period that history will record as a policy revolution in British Columbia, he acted.

CHAPTER 2

The Rise of Climate Action
The Role of Gordon Campbell

President John F. Kennedy often told the story of the aged Marshal Lyautey of France debating with his gardener the wisdom of planting a certain tree.
  "It will not bloom," the gardener argued, "for decades."
  "Then," said the marshal, "plant it this afternoon."
       JAMES GUSTAVE SPETH, *Red Sky at Morning*

AT ITS HEART, HUMAN-MADE CLIMATE CHANGE IS AN INTERgenerational issue: incremental, slow, even stealthy—and, above all, long-lasting. Those adjectives describe a challenge that has stymied politicians for decades. Why act on something that will not have obvious societal payouts for generations? Why risk actions that might be unpopular, when benefits will be realized well past the current electoral cycle? Credit for taking bold steps that will improve future society, or the environment, or the economy, or all three at once, can be hard-won. And yet, every so often a politician comes along who puts principle before popularity, who takes the right action and plants a figurative tree: not because it's popular, but because it's the right thing to do.

Gordon Muir Campbell, thirty-fifth mayor of Vancouver (1986–93) and thirty-fourth premier of British Columbia (2001–11), was one such politician. Love him or hate him, Campbell designed a policy that will be remembered for decades for its bold response to climate change. How he got there, and what he did, is a story that stands apart in the murky, often disappointing and sometimes malfeasant world of climate politics.

Campbell was no political neophyte in adopting environmental sustainability. His commitment to the concept was clear from the beginning of his career in civic politics, or even before. When asked when he became personally interested in climate change, he said, "This will come as a surprise, but it was a long time before I was even running for office."[1]

It didn't take long for him to express that personal interest after becoming Vancouver mayor in 1986. A developer previously in his career, he valued long-term urban and regional planning. But his hands were tied. A few years earlier, Premier Bill Bennett imposed a fiscal bloodletting on the province. Inspired by the neo-conservative economist Milton Friedman, Bennett and his team slashed spending and governmental services in 1983. Included in that so-called "restraint program" was the cancellation of all official regional plans and "the removal of regional planning as a function of regional districts."[2]

Campbell could not let that stand. In early 1988, just over a year after becoming mayor, he assumed the chair of the development services committee of the Greater Vancouver Regional District (GVRD), a body formed to "oversee the resurrection of regional planning."[3] He was just thirty-nine years old. Ambitious and energetic, and with growing awareness of global-scale environmental issues, Campbell saw his role as committee chair as promoting the transformation of BC's Lower Mainland into one of the world's most livable and sustainable urban settings. The region needed guidance in avoiding the sprawl rapidly metastasizing around so many major North American cities.

Together with three talented others—Walter Hardwick, an urban geographer at UBC; Ken Cameron, a senior planner with the GVRD; and Judy Kirk, a GVRD communications manager—in 1989 Mayor Campbell launched the highly consultative Choosing Our Future process, the broad scope of which was to include "social inclusion, urban design, justice, transportation and the environment."[4] That effort culminated in a needed road map, *Creating Our Future*, a document released in September 1990 that laid out sustainability targets.[5] In the 2007 book *City Making in Paradise*, Campbell's input to the report is described as yielding "an inspirational value-driven agenda for the region."[6] (That book was co-written by Mike Harcourt, who preceded Campbell as both mayor and premier.) After the document had been approved by the GVRD board, Ken Cameron singled out Campbell for specific praise, noting that his leadership and ability to partner "restored the region's ability to think ahead."[7] Those words, in 1990, foreshadowed what Campbell was to deliver some two decades later.

The focus of *Creating Our Future* was firmly regional. But as it was being developed, international environmental concern was mounting. The future looked increasingly perilous. Atmospheric greenhouse gas concentrations were escalating exponentially, temperatures were rising, and a growing chorus was demanding action to curb emissions.

In June 1988, the World Meteorological Organization, the United Nations Environment Programme and the Government of Canada co-hosted, in Toronto, arguably the most important international environment conference of the decade. Entitled "Our Changing Atmosphere: Implications for Global Security," it yielded a consensus that stood as a clarion call: "Humanity is conducting an unintended, uncontrolled, globally pervasive experiment, whose ultimate consequences could be second only to global nuclear war."[8]

Like most Canadians, Campbell—in the mayor's chair in Vancouver—heard that call. He and his council responded, commissioning a seven-member "task force on atmospheric change," whose mandate was to gather public input and recommend specific actions.[9] The task force released its landmark study, *Clouds of Change*, in mid-1990. Campbell singles it out as a signpost of his interest in environmental public policy.

Just like *Creating Our Future*, *Clouds of Change* was inspirational, laying out a comprehensive suite of recommendations designed to reinforce long-term sustainability. One key clause specified that carbon dioxide emissions should be reduced 20 per cent by 2005 relative to 1988 levels, in part by establishing a regional carbon tax, instituting road pricing, reducing the use of gasoline and diesel in vehicles, exploring the use of alternative fuels, and supporting non-auto transportation opportunities. The carbon tax would fund some of these opportunities. The authors of *Clouds of Change* knew they were asking a lot, noting: "While we recognize that some of these recommendations may seem ambitious, we believe that to do less would be to shirk the responsibility of our generation."

Campbell implicitly understood that responsibility. But it was to be almost two decades later before he was able—or more accurately, before he chose—to act on it. He stepped down as mayor in 1993, ran for the leadership of the BC Liberal Party, and won a three-Gordon race in which he defeated Gordon Wilson, the incumbent leader, and former MP Gordon Gibson. He served as BC's leader of the Opposition until 2001, when his party won seventy-seven of seventy-nine seats—the most lopsided electoral victory in the history of the province.

With that extraordinary majority, Campbell had the opportunity then, in 2001, to address "the responsibility of our generation."

But he didn't.

The reasons why wrap around the very core of his personal political ideology. When asked specifically what he saw then as the balance between environment and economy, he replies, "The environment is fundamental," adding: "I don't agree that you've got to be for the environment or for the economy. I think you've got to be for both and be smart about it." But following his election as premier, the environment took a back seat. Instead, Campbell's main goal was to focus on the economy. People may value the environment, he says, but their immediate priorities will always be their children, their family and their immediate future. "If you don't build a strong economy," he reasons, "people are going to say, 'I don't care about the environment right now—I've got to make sure my kids are eating.'"

Martyn Brown, Campbell's chief of staff as premier, shared that view in 2001. In those years, climate change "was so far removed from our agenda, it wasn't mentionable."[10] Now a retired political commentator and writer, Brown was the primary author of the 2000 BC Liberal platform, *A New Era for British Columbia*.[11] That document is riddled with clauses like "Our plan is aimed at kick-starting the economy in every sector" and "liberating our economy and minimizing undue government intervention in people's lives." Finding phrases like *climate change* or *global warming* in the *New Era* text is a fool's errand. They aren't there.

In those early years of his premiership, Campbell failed to build on his decade-old reputation as a sustainability champion, earned while serving as mayor. Martyn Brown says: "Under Treasury Board's direction, the government largely gutted the Parks budget in its first couple of years. Despite then-environment minister Joyce Murray's passionate appeals to the 'red meat' crowd in cabinet, that first administration really didn't do very much at all for air and water protection and all sorts of other things: species at risk, wildlife protection." As

Brown puts it, in those days, Campbell's "one thing across the board was doing the right thing in a way for the economy that he could feel good about."

Indeed, it was to be several more years before *climate change* became a central phrase in British Columbian politics, several years during which carbon dioxide concentrations increased another 3.5 per cent, global average temperature rose another tenth of a degree, mountain pine beetles continued to paint BC's Interior forests various hues of orange and red, and the world's climate scientists grew increasingly anxious about the future.

Premier Campbell was well aware of that growing anxiety, in part because he travelled the province and witnessed the havoc being wreaked on Interior forests by the pine beetle. "I was aware of its economic impact," he says, while denying that it was the key driver that spurred him to take action. Instead, he describes the infestation of the early to mid-2000s as a proof point. "It was a critical example of why we had to act on climate change." He would tell those who disagreed to look at what was happening in the forests and with the average temperatures. "We wouldn't have the same epidemiology of the pine beetle in the forests if we didn't have climate change taking place," he says.

Ask Gordon Campbell's friends or colleagues about his reading habits, and they will tell you he reads "everything." As premier, he always had a stack of books at his side, the titles of which often addressed sustainability issues, global change, and environmental integrity in the face of growth, both human and industrial. One in particular, in the fall of 2006, had a profound impact: *Red Sky at Morning*, by Yale professor James Gustave Speth.[12]

Speth was prescient in a way that resonated with Campbell. Speth wanted his book to be a wake-up call for everyone "who may believe that all the international negotiations, treaties and other agreements of the past two decades have prepared us to deal with global environmental threats. They haven't."[13] For a new design to happen, argued Speth, "civil society must take the helm." He went on to point out that environment versus economy—a common conservative construct, typically camouflaged by repeating the word *jobs*—was a false dichotomy. That, too, resonated with Campbell. Speth sought a world "in which market forces are harnessed to environmental ends, *particularly by making prices reflect the full environmental costs.*"[14] That concept appealed directly to Campbell's free-market, conservative roots. It also offered a political advantage: a true conservative understands what it means to conserve. Reflecting on his thinking both then and now, he says, "If you are conservative economically, you can wear the environment if you want."

Red Sky at Morning laid down a serious challenge to civil society: take the helm, and at the same time couple the free market with environmental stewardship. Campbell was increasingly in a position to do both, while being acutely aware that it wouldn't be easy: "One of my theories of public life is that if you try to take a giant leap you are often going to fall into a chasm. But if you actually bring people with you and tell them what you are doing and why you are doing it and encourage them to be part of it, then you can actually make some progress."

When he was re-elected in 2005, Campbell's literary influences critically shaped his thinking as he continued as premier. While on vacation in the winter of 2006–07, he read Tim Flannery's international bestseller *The Weathermakers*, as well as *Heat* by George Monbiot. Both focused on climate science: why global warming is occurring, and what we can do about it. A biologist, paleontologist, television personality and prolific author with a position at the University of Melbourne, Flannery is Australia's equivalent to David Suzuki. His commanding knowledge and skill with the pen match that of Monbiot, an outspoken and exceptionally well-regarded journalist with the UK's *Guardian* newspaper. Both authors are committed environmentalists, but neither is a doom-and-gloomer. Rather, both offer solutions to climatic peril, and like Speth, they lay out a societal challenge, pointing out that energy provision and delivery systems must be reconstructed globally. Their writings resonated with Campbell.

But as the saying goes, all politics are local. While the global scale is key, Campbell had a more specific focus: How can we best take constructive action at the level of a single Canadian province, or even locally?

University of Victoria professor Michael M'Gonigle and Justine Starke addressed that question in their 2006 book *Planet U: Sustaining the World, Reinventing the University*. UBC professor emeritus William Rees, co-founder of the term *ecological footprint*, described *Planet U* in a review at the time as a call to arms: "M'Gonigle and Starke provide the rationale and inspiration to break rank from the growth and globalization mainstream ... Only by reinventing itself can the university hope to become society's champion for the locally rooted global sustainability that is the quest of peoples and communities everywhere."[15] Upon reading the book, Campbell said it was a "lighter of the fuse." It led him to wonder: If becoming a sustainability champion is such a brilliant idea, why wasn't the University of Victoria leading it? He addressed it directly a year or two later, enshrining in BC legislation a requirement to practice sustainability at the local institutional scale.

A second fuse was ignited when Campbell travelled to Beijing in the fall of 2006, where he encountered opaque white air—air that you could taste. There was so much pollution that he could barely see the other side of a street. He describes it as an "aha moment" that led him to the idea that "the little things that each of us does has a huge impact on the overall environment, as opposed to always talking about 'the big.' I thought we should be talking about 'the little.'" Upon returning to BC, he raised the issue with caucus, who agreed to take initiative. In stark contrast to the new federal Conservative government, Campbell says, "I thought it was critically important to say that we were actually going to do this and we are going to use well-informed public policy based on the science." It led his philosophically conservative government to take concerted action.

Events south of the border, led by another conservative, helped make the case. In August 2006, California's Republican governor Arnold Schwarzenegger had signed into law Assembly Bill 32, the Global Warming Solutions Act, that required the state to reduce its greenhouse gas emissions to 1990 levels by 2020.[16] Martyn Brown acknowledged that AB 32 showed that "there was something the world could do." And when asked about the first time he met Governor Schwarzenegger, Campbell jokes, "Well, I was in an action film and a lot of people didn't notice me there." *Terminator* humour aside, the truth is that the premier called the governor in December 2006. They had a meeting of the minds. In Campbell's words, they "shared similar frustrations with our federal governments." Brown says that Campbell saw "a great opportunity for them to put their heads together—for [Campbell] to be profiled through Arnold Schwarzenegger's leadership, but also for California to be partnering with a Canadian jurisdiction because Washington [DC] was doing too little on climate policy."

At that point, California was clearly the continental leader on climate action. But it didn't take long for BC to join them. Three months after Campbell's initial call with Schwarzenegger, the 2007 provincial Throne Speech was delivered by Lieutenant-Governor Iona Campagnolo. Martyn Brown wrote much of the text of the forty-two-page document.[17] The timing for the Campbell government was right. Brown says, "The world [had] started thinking about climate change as a public issue," citing Al Gore's film *An Inconvenient Truth* as a pivotal moment. Schwarzenegger, a conservative, had acted in California. In BC, as the Throne Speech described, the economy was on track, a triple-A credit rating was regained, the province led the country in job growth, and its health system was ranked as the best in Canada.

And there was something else. Brown notes that the government was looking for new fields of endeavour. "Climate action was reflective of the government's state of mind—the premier's state of mind—about what to do next. What's the point of this job if not to change the world for the better?"

Looking back, Campbell echoes Brown: "It was meant to be a strong provincial initiative that would give people across the country the opportunity to say, yes, let's watch what they are doing in BC and maybe we can move forward." He didn't see it as right-wing or left-wing. It was simply the right thing to do.

The 2007 Throne Speech was remarkably forward-looking, posing a question in its opening paragraphs that immediately captured the essence of the challenge facing humanity: "What can we do today to secure the future for our children and grandchildren?" It went on to ask another: "Will we have the courage to tackle difficult problems that have no easy solution?" And it rhetorically responded: "To these questions your government answers—yes." There was to be no waffling, no partisan machinations. Martyn Brown's text made that clear: "This is a time for partnership not partisanship, for boldness not trepidation, for action not procrastination."

The legislative road map of the next eighteen months differed from California's pathway. Although BC "learned a lot" from California, according to Brown, politicians in the state "didn't believe in a carbon tax." There were powerful political reasons why. Since at least the days of Ronald Reagan, the word *tax* had become something to be avoided in US politics, unless accompanied with the word *lower*. Indeed, it has been said that if there were a dictionary of the modern American language, *tax* would be defined as "poison." That semantic perversion—a neo-conservative debasement of what taxes are, namely the cost of constructing and operating a civil society—never existed in Gordon Campbell's mind. When asked if California's legislation helped to shape BC's climate-action plans, Campbell replies, "No, we decided we'd take an independent path." *Tax* was to become a keyword in the steps that British Columbia was planning just over the horizon.

Throne speeches are typically big-picture documents that lay out, in figurative large print, where a government intends to go. The 2007 speech was different from most: it added fine print in describing more than fifteen pages of initiatives—over one-third of the entire speech—in its "Environmental Leadership" section. In making the case for climate action, it stated that no priority "is more important than the critical problem of global warming and climate change," adding: "The science is clear. It leaves no room for procrastination. Global warming is real." And in a nod to urgency, it called for serious engagement: "The more timid our response is, the harsher the consequences will be."

The four dozen clauses that followed eschewed timidity, stating an aim to reduce BC's greenhouse gas emissions by at least 33 per cent below current levels by 2020: "All new and existing electricity produced in BC will be required to have net-zero greenhouse gas emissions by 2016." It announced that "effective immediately, British Columbia will become the first jurisdiction in North America, if not the world, to require 100 per cent carbon sequestration for any coal-fired project." The new energy plan included a requirement for zero natural-gas flaring. Tailpipe emission standards for all vehicles sold in BC would be phased in by 2016, and a low-carbon fuel standard was to be established. A new BC Green Building Code was to be developed. The mountain pine beetle made an appearance, with an announcement that infested trees—wood chips and other wood waste—would be used to create new energy and clean power. And in a nod to *Planet U* and Campbell's earlier musings about the University of Victoria, the speech announced that options would be explored to make "the government of British Columbia carbon-neutral by 2010." "Government" in this case was to include all public universities.

All of these actions were being proposed by a conservative administration.

The phrase *carbon tax* is missing from the 2007 speech. Martyn Brown points out that wasn't overt.[18] The intent was there, he noted, just couched in different words that were intentionally broad. He says phrases like *market forces* or *market pricing* were shorthand for an anticipated cap-and-trade system in addition to, or parallel with, a tax.

The philosophical underpinnings of a carbon tax were also there. Gordon Campbell believes strongly in *how* we make responsible choices, both personally and collectively. His philosophy—and make no mistake, it is his—was captured by several sentences in the speech that pertain to climate action: "What each of us does matters. What everyone does matters." And later, by saying that the government believes "our tax system should encourage responsible actions and individual choices."

That notion—changing the tax system to favour wiser environmental choices—wasn't met with universal applause, even within Campbell's government. According to Martyn Brown, some caucus members thought "we were being way too radical and aggressive for the good of our economy," complaining that the proposed climate actions "wouldn't make one iota of difference to the planet." Along the same line, members of the business community quietly asked,

"What the hell does it matter? BC's GHG emissions are just a drop in the bucket."

That argument—so often trotted out in climate-denial and right-wing circles—is remarkably specious. Brown had anticipated it in the Throne Speech, writing:

> The argument that British Columbia's mitigation efforts are, in global terms, too minuscule to matter misses the point.
>
> Every molecule of carbon dioxide released into our atmosphere by human activities matters. It hangs there for decades or even centuries, and adds to the accumulated burden of global warming on our planet.
>
> The benefit of our actions is not negated by the actions of others who add to the problem. They are cumulatively beneficial, globally significant and scientifically discernible. They contribute to the efforts being taken by growing legions of people around the world who are acting today to prevent the problem from becoming even worse.
>
> We cannot be paralyzed into inaction by the scale of the task at hand. Rather, we will act now to make a real difference.

In reflecting back to those heady days, Campbell remembers how strongly he and Schwarzenegger felt about making that real difference. "We both agreed that we were going to take some actions that would lead locally or regionally, and hopefully the regional example would set something for the national governments to follow, both in Canada and the US. Both federal governments were very slow to pick up any of the challenges, even the most straightforward challenges." British Columbia joined the Western Climate Initiative (WCI) and took a number of steps to outline several policies that would have a cumulative impact.[19]

Whereas joining the WCI was one of those steps, there were caveats. British Columbia was determined to take an independent climate-action path, at least in those early stages. That route was to be a direct reflection of Campbell's personal philosophy: allowing citizens to exercise choice. The centrepiece of the WCI was to be a cap-and-trade program wherein large emitters in the member jurisdictions would be required annually to have permits or credits to pollute. At the outset, existing emitters would be given free credits to match their emissions—one credit per tonne of carbon dioxide emitted—but in the following years state or provincial governments would set progressively lower allowable emissions levels, i.e., caps. Big polluters that could not meet their caps would be required to purchase credits at auctions, while emitters that became more efficient could sell "excess" credits. In that way, a carbon market would be established, and emissions would be forced to decline toward collective limits defined by government. That, at least, was the theory.

Textbook theory gets complicated and perhaps even perverted when entitlements enter the equation. Cap-and-trade schemes are fertile grounds for lobbyists, who argue that their particular client could face economic hardship and deserves an exemption—either temporary or permanent—from the scheme. Moreover, issuing too many credits at the outset, as happened when the European Union created its Emissions Trading System, reduces demand at the market level, keeping prices low. Emitters that need to buy credits then have little fiscal incentive to invest in low-emissions pathways. Climate action gets pushed into the background.

Campbell was well aware of these pitfalls and recognized that cap-and-trade wasn't an ideal fit for British Columbia, at least in 2007. Unlike California with its coal-fired power plants, hydropower-rich BC didn't have many large-scale, point-source emitters. Moreover, cap-and-trade programs do not directly influence the carbon-emitting practices of individuals, which was a core consideration for Campbell.

Given those issues, a revenue-neutral carbon tax was his preferred step. He was willing to pursue cap-and-trade primarily because California was. The question of exemptions was always present. "Cap-and-trade was as much a fight about how you didn't do something as how you did," he says. "I would say to everyone today: don't spend time on cap-and-trade, because people are going to spend their time arguing ... trying to game the system. And the nice thing about the revenue-neutral carbon tax is that it was a straightforward, pretty elegant way of doing something that was important." It also fit well with Speth's entreaty in *Red Sky at Morning* to couple the free market with environmental stewardship.

Although the 2007 Throne Speech did not contain the phrase *carbon tax*, it did provide foreshadowing: "Over the next year, the province will consider the range of possibilities aimed at encouraging personal choices that are environmentally responsible. It will look for new ways to encourage overall tax savings through shifts in behaviour that reduce carbon consumption." Over that next year, Campbell's government talked to "various business groups from twelve or thirteen different sectors of the economy," he says. "We had groups come in and speak to the cabinet committee on climate," a body he chaired. When the cap-and-trade option was raised, "we were told by virtually the entire business community that they wanted no free riders, they didn't want any exemptions, [and] they wanted to have predictability over the long term." That community got its wish.

The 2007 Throne Speech will be recorded as a seminal moment in British Columbia's political history. But while it opened the door to serious climate action, it also posed arguably the biggest challenge any government anywhere can tackle: changing human behaviour.

The government recognized both that imperative and its inherent difficulty in the closing words of the speech:

Let us test our limits and give our grandchildren the gift of a better province, a better country, and a better world.

Almost exactly one year later, the 2008 Throne Speech boldly built on its predecessor, announcing that the government was not hesitating to follow through: "British Columbians are taking decisive action on climate change," it stated.[20] And in echoing the speech of a year earlier, it declared: "Your government will be carbon neutral by 2010."

Campbell was adamant that the government itself, and not just individuals, had to set an example, saying: "It starts with government. We had people say we can't do carbon-neutral government. And we said, 'Well, we're going to.'" And in an implied challenge to jurisdictions elsewhere, the speech announced to the world that British Columbia refused to stand by, that waiting for others to act only compounds the problem: "Taking refuge in the status quo because others refuse to change is not an answer. It's avoiding responsibility and being generationally selfish." The speech made it clear that the necessary policy details were to come in short order, including imminent release of a climate-action plan "founded on personal responsibility, sound science and economic reality."[21]

The pronoun *we* is prominent when Campbell reflects on those years. "I want to make it clear that I had lots of input from people all around," he says. He eschews the personal pronoun *I*, noting: "I had a couple of political colleagues who would always stand up and say, 'I did this and I did this and I did this.' I hated that. Because even if you were a minister and you think you'd launched something, you didn't do it—you did it with your caucus. You did it with these people who said yes."

He is generous with his praise when discussing those who helped shaped the climate-action plan, singling out the trio that he considers its principal architects: Jessica McDonald, Graham Whitmarsh and Martyn Brown. As deputy minister to the premier, McDonald held the most senior public service position in the government from 2005 to 2009. Campbell describes her role in shaping the plan as "pivotal." Whitmarsh was head of the young Climate Action Secretariat. A savvy former nuclear engineer who served in the Royal Navy aboard nuclear-powered submarines, Whitmarsh once quipped, "I slept for years with my head a yard from a reactor." He knew his stuff. Campbell praises his technical prowess in particular.

But it is Brown—chief of staff and principal author of both the 2007 and 2008 Throne speeches—who Campbell particularly commends as "brilliant." The two had "a very synergistic, open, honest, direct relationship," says the former premier, describing his right-hand man as an intellectual driver who "was inspired by the ideas and he inspired us back, so I would say he was a critically important component of our climate strategy."

But in looking back at those heady days of climate announcements, there remains the niggling puzzle of the two missing words: *carbon tax*, the undeniable core of British Columbia's climate efforts. And yet that specific phrase is absent from *both* Throne speeches, not just the 2007 document. Should anything be read into that omission? Was Premier Campbell paying heed to the perverse American view that *tax* is a toxic word? According to Martyn Brown, the government simply preferred to continue using broader language, like *market forces*, in part to keep its options open. Brown says, "It's fair to say we avoided the phrase *carbon tax* in that budget year that introduced it." The phrase had been bandied about for a year prior. But seven days after the 2008 speech was delivered, two precise words—*carbon* and *tax*—were first purposely juxtaposed by a cabinet minister on the floor of the House.

On February 19, 2008, a cloudy but rain-free day in Victoria, Finance Minister Carole Taylor rose in the House to present the provincial budget.

She was wearing green shoes.

Taylor read three paragraphs that were to make history:[22]

> *With this budget, we are introducing a major shift in the way we levy taxes. Effective July 1st, we intend to put a price on carbon-emitting fuels in BC. This carbon tax will be entirely revenue neutral, meaning every dollar raised will be returned to the people of BC in the form of lower taxes. Furthermore, that commitment will be enshrined in law. And we will provide additional support for lower-income British Columbians.*
>
> *The carbon tax will start at a relatively low rate to give us all—businesses and individuals—time to adjust. As it rises, it will underline the fact that there is, indeed, a cost attached to generating greenhouse gases.*
>
> *Leading economists and scientists agree: Seeing that cost, and making it real, will give us new incentives to change the habits that created global warming in the first place.*

Taylor knew how momentous this was. Asked almost ten years later if she'd felt excited or even apprehensive at that moment, she replies, "No. I felt this was the right thing to do. And as a politician, you can waver and be worried and concerned as you go along through various decisions, but if you land in a place that you can say, 'This is the right thing to do,' there's a calmness to that. I felt very good about making history, really, and doing the right thing for British

Columbia." Chuckling, she adds that there were "lots of frowns on the other side of the House."[23]

Few knew how much effort had gone into crafting the legislation that supported those three paragraphs in the budget speech. Taylor worked almost in secrecy for a full year with Premier Campbell and a small coterie of senior officials at the Ministry of Finance, in the wake of the 2007 Throne Speech. No one else was involved. She lists three or four basic questions with which they wrestled: Is a carbon tax something we should do, or shouldn't do? If we did it, what would the model be? What could we do to make it successful? "It was an intense, fantastic time for me and my finance staff," she says.

Keeping the circle small was crucial. As the planning took shape, they kept the lid on. "It was very isolated," she says. "You cannot talk to anybody else about [the tax] because it always moves the market. So we couldn't even say if we were going to do a carbon tax." The seventy-seven-member BC Liberal caucus was purposely kept in the dark. "We did not tell caucus about the carbon tax until the day I delivered the budget," she says. Why so strict? "Because someone would leak it for sure, and then the market would go crazy."

Taylor suggests that as an outsider to politics—her previous career was as a journalist—she brought a different perspective to the discussions. In her mind, the key was to build the political and social rationale, observing that "a fine academic policy is a useless policy if you can't get acceptance." But she and Campbell ran into an immediate problem and she resorted to camouflage to circumvent it. "Right from the start, I said, 'You have to be able to talk to people about this and yet I can't talk to them.'" Taylor's ministry looked at about a dozen possible permutations for a solution, soliciting opinions from business and environmental groups.

But it was all a disguise, for she and Campbell were already thinking that a revenue-neutral carbon tax was the way to go. As she puts it now, "The question then was: How can you design it in a way that people will accept, that business will accept, and environmentalists will accept?" She and her team looked around the world for a suitable model. They found none. "And so we thought, 'Okay, we're starting from greenfield here. Let's build it.'"

What they went on to build—British Columbia's Carbon Tax Act—became one of the most applauded pieces of legislation ever signed into law in the arena of legislative environmental responsibility in North America. And at its philosophical heart sits a quartet of words that the lieutenant-governor had read in the House a week before Carole Taylor slipped on her green shoes: "You choose. You save."

Those four spare syllables summarize much of Gordon Campbell's conservative political philosophy. Reflecting back, he says, "One of the things we were trying to do with climate change is to say: we'll create an environment that allows you to make choices. I'm not going to tell you to have a big car or a little car, but if you buy a big car it's going to cost you more because there are more costs to the environment when you drive it." The notion of tax *shifting* was deeply embedded in that position. "Revenue neutrality was a critical part of what we were doing," Campbell says, contrasting it with Liberal Leader Stéphane Dion's ill-fated Green Shift policy in the 2008 federal election campaign.

Just before BC's carbon tax was to kick in on July 1, 2008, federal Opposition Leader Stéphane Dion announced that, should he become prime minister, his government would legislate the Liberal Green Shift, a new national program to combat global warming. Taxes would be levied on greenhouse gas emissions, and the revenue used to cut federal income taxes, enhance child benefits and strengthen support for seniors and low-income families. Ostensibly revenue-neutral, Dion's program differed from BC's initiative in one politically fatal way: when imposing new taxes on energy use, the nation of Canada could not be treated like a province.

That distinction was nowhere truer than in the West. Alberta and Saskatchewan were heavy emitters, owing largely to coal-fired electricity grids and oil production. "Who should be assigned the carbon footprint [and thus the carbon tax] for producing oil?" asked Queen's University economists Thomas Courchene and John Allan, according to a post-election report by Janice MacKinnon.[24] MacKinnon, a former Saskatchewan finance minister, noted that both provinces "would be hit very hard by a carbon tax." Moreover, their relatively small populations would mean that neither would benefit much from the social programs. "The carbon tax would have the effect of transferring revenue from Alberta and Saskatchewan to be used in other regions," she contended, implying that revenue neutrality would not prevail in those two jurisdictions.

That consequence—unaddressed regional inequities in tax receipts and revenue allocation—sank Dion's Green Shift almost as soon as it was announced. Conservative PM Stephen Harper successfully labelled it a tax grab that would alienate the West, a charge that Dion was ineffective at countering. In the next election, popular support for the Liberals nationally fell to below 27 per cent, the lowest in the history of the party, and only two Liberals (against forty-nine Conservatives) won seats on the Prairie provinces. Stephen Harper remained prime minister, and Dion resigned as Liberal leader six days later.

Campbell describes Dion's plan as a failure. "It was a revenue grab, as opposed to a change in policy that said 'We're going to tax emissions and not tax something else.'" When asked specifically why he felt so strongly about revenue neutrality, Campbell zeroes in on personal choice: taxing emissions while reducing income taxes "so that you can decide what you are going to do." If you use more carbon, said

Campbell, you're going to pay a higher tax; if you generate less carbon, you are going to pay less tax and have more money in your pocket. In his mind, it was all about choice and responsibility: big car or small car, air-source heat pump or natural-gas furnace—you choose.

Martyn Brown disagreed with his boss; he wasn't convinced that revenue neutrality was that important: "My argument there was 'Will anybody see the tax benefits?'" He's mellowed on that today, noting that he thought it "would be more problematic than it turned out to be." But some BC residents—maybe even most—shared Brown's concern. Revenue neutrality was an alien concept to them. Paying more at the gas pump hit their wallets directly, but personal income tax reductions were essentially invisible.

Near-invisibility of lower income tax aside, Campbell's climate-action legacy is secure. He recalls the words of an environmental activist who introduced him at an event in London, years after he'd left the premier's office and became Canada's high commissioner to the UK: "This is the man who was the premier when the most effective climate strategy in the world was put in place." While Campbell is the first to admit that he doesn't deserve the sole credit, he does note that being introduced with those words "sure made me feel good," adding: "I think Canadians feel good doing the right thing for the right reasons in the right way." Ever the pragmatist, he admits that the actions his government took in 2007 and 2008 weren't perfect, but asks, "What is perfect in this human world? They were sure good steps in the right direction. When we look back generationally we can say: we did what we needed to do. I think there's more we need to do. I think we should get over this discussion about economy or environment. We should recognize that you can do both if you are just smart about it and change the way you think and the way you do things."

Amen.

CHAPTER 2

EPILOGUE: *Gordon Campbell resigned as premier on March 14, 2011, five months after announcing he would step down. The primary reason for his resignation had nothing to do with his environmental record. Rather, he was accused of misleading the public when his government moved in mid-2010 to combine the 7 per cent provincial sales tax with the 5 per cent federal goods and services tax to produce a 12 per cent "harmonized sales tax," something Campbell had said in the 2009 election campaign that he had no intention of doing.*

When Campbell resigned, British Columbia lost a legislative climate champion.

By addressing the climate challenge, Gordon Campbell and his team planted Marshal Lyautey's tree in British Columbian soil. Its roots had grown. Its trunk was solid. Small limbs had the potential to thicken and branch out. But as it turned out, Campbell's successor as premier, Christy Clark, did not have a green thumb. Under her watch, the tree was neglected. It did not die. It just did not bloom.

CHAPTER 3

British Columbia's Carbon Tax

"A template for the world"

"**M**R. MINISTER, I WANT YOU TO KNOW: YOUR CARBON TAX IS A template for the world."

So said King's College professor Paul Ekins in his gentle London accent, at a private meeting in January 2010 with British Columbia's Minister of State for Climate Action John Yap. Yap had been appointed by Gordon Campbell to oversee delivery of the climate actions laid out in the Throne speeches of 2007 and 2008. He didn't normally hear that a key policy his government had implemented was a model for the rest of the planet. He beamed.

One of the world's leading environmental economists, Ekins had been appointed by British Prime Minister Gordon Brown in 2007 to direct the UK's newly created Green Fiscal Commission. Brown was a democratic socialist who would have sat across the aisle from any government that included Gordon Campbell. But Brown recognized that the need to tackle the climate challenge didn't depend on one's political stripe. He charged the all-party, multi-sector Green Fiscal Commission with finding optimum ways to shift the British tax burden onto pollution and away from income or employment taxes.

The commission had submitted its report in October 2009, but the British government had yet to act.[1] Ekins came to British

Columbia three months later to learn more about its carbon-pricing approach, thinking it could be a model for the UK. After four days of speaking to academics, the public, the business community and government officials in Vancouver and Victoria, he was impressed. He had a template.

British Columbia's carbon tax took effect on July 1, 2008. It was remarkably simple. When Finance Minister Carole Taylor stood in the House five months earlier to introduce it, she touched on the key design elements: revenue-neutral, thus a tax shift, not a revenue generator; a low tax rate at the start and an upward ramp over four additional years, thus a signal to markets and individuals that polluting was going to cost progressively more as time marched on; and direct provision of support for low-income British Columbians, thus fair.

The details are important.[2] First, the tax was universal. Both Gordon Campbell and Carole Taylor insisted that it apply to everyone in the province. Taylor is firm: "It had to be a system that didn't have exceptions all over the place, because I had been through that with other taxes." You can't buckle, she asserts. If you say, "Oh well, we'll except the teachers for this or we'll except the builders for this," then "all of a sudden it falls apart." Campbell is of the same mind, noting that the BC business community "didn't want any exemptions, [and] they wanted to have predictability over the long-term. All of those things were part of what was included and tailored into the carbon tax."

The tax was very broad, and applied to all emissions generated by burning tires or peat or combusting fossil fuels purchased in British Columbia. There were only two exceptions to the latter: aviation fuel used for out-of-province flights, and fuel used in cruise ships that sail to other jurisdictions, since most of the combustion of such fuels would occur outside BC's borders. Anyone catching a flight to

Maui from Vancouver or boarding a ship in Victoria to sail down the coast to California wouldn't pay the carbon tax. But if you flew from Kelowna to Prince George or Prince Rupert or Fort St. John, or took a ferry from Comox to Powell River, the tax was built into your fare.

Second, the tax was initially trivial, set at $10 per tonne of carbon dioxide emitted. That translates to just 2.34 cents on a litre of regular gasoline and 2.69 cents for diesel (Table 1), less than the daily or weekly ups and downs at the pumps. When Gordon Campbell is asked why the initial rate was set at $10 per tonne and not some number lower or higher, he says, "It seemed like a good start." Martyn Brown says they didn't do any polling on it, but that setting the initial tariff "was more of an assessment about what the political blowback would be."

Ten dollars a tonne *was* just a start, for the tax was scheduled to rise on July 1 of each of the next four years, to $30 per tonne emitted on or after July 1, 2012. Five dollars per tonne per year equates to about a 1.2-cent annual increase in the cost of a litre of regular gasoline, a rate that Campbell and his government judged to be acceptable to average consumers. Carole Taylor notes that the government purposely started low. "Environmentalists were not happy with that," she points out. "They said, 'Well that's not going to make any difference.' [But] in my mind, I had to get the principle accepted. If the principle is accepted, in years to come politicians can adjust it."

That principle—a low starting point and a slow upward ramp spread over several years—was designed to encourage gradual adjustment in fossil fuel use both by industry and private consumers, while minimizing economic dislocation. Taylor says the legislated $5-per-year rise satisfied the business community's need for predictability: "Business very often will adapt quite well to whatever the rules are, but you've got to *tell* them what the rules are. No discussion, no debate. Write it into your business plans."

TABLE 1. British Columbia carbon tax rates for selected fuels as of July 1, 2008, and as of July 1, 2012, following four years of successive increases. The 2008 and 2012 rates are calculated on the basis of $10 and $30 (CAD) per tonne of CO_2 released via combustion, respectively. Note that higher-carbon fuels are taxed at a higher rate. This design feature ensures that CO_2 emissions produced during combustion of each fuel are equivalently priced. Other combusted fuels, including shredded and whole tires as well as peat, are also taxed. *Data from Schedule 1 of the Carbon Tax Act of British Columbia (2008).*

Fuel	Units	2008 Tax Rate	2012 Tax Rate
Gasoline	¢/litre	2.34	6.67
Diesel	¢/litre	2.69	7.67
Natural gas	¢/cubic metre	1.90	5.70
Propane	¢/litre	1.54	4.62
Coal, high heat value	$/tonne	20.77	62.31
Coal, low heat value	$/tonne	17.77	53.31

Another factor more directly confronted political reality. Legislating the increases over a number of years "was a way not to have the debate every single year," she says.

Of all the design features of the tax, the third was the most important. It sat at the very core: the mandate to be revenue-neutral. According to Gordon Campbell, that was a "clear policy direction" laid down by the cabinet committee on climate change, which he chaired. He says Taylor spearheaded the revenue-neutrality objective, and that she was "active, engaged and creative." She recommended that every penny of revenue be returned by law to the residents of British Columbia through reductions in personal, small business and corporate income taxes—and some direct fiscal transfers. "Everybody is going to get hit by this tax," she said, "so everybody should be helped by this tax."

Although British Columbians didn't know it until the budget came down on February 19, 2008, they were to be given a retroactive leg-up on the carbon tax. That's because personal income tax rates in BC were about to be reduced by 5 per cent on taxable incomes up to $70,000. That decline in provincial tax took effect as of January 1, 2008, six months *before* the carbon tax was to kick in on July 1. "It gave BC the lowest income tax up to $110,000 anywhere in the country," according to Taylor. It was, in her words, "a big change."

Corporate income taxes were also reduced—a bit later on, and in steps. General corporate income tax declined from 12 per cent on July 1, 2008, to 10 per cent on January 1, 2011, which took it to a level among the lowest in the OECD countries. Proportionally, small businesses—those with annual income up to $400,000—benefited even more, with the tax rate falling from 3.5 to 2.5 per cent on December 1, 2008, a net reduction of nearly 29 per cent.

But tax reductions only impact those individuals or companies that have income. Taylor, who cites her "profound sense of social justice," was concerned about those who didn't. "I worried about those who were very low-income and middle-income, because people who are well-off, or businesses, will take care of themselves," she says. The immediate challenge became helping "those folks who would still have to pay extra for gas for their pickup truck and they'd have to pay for their heat up north." When she and her team started playing with models, she says it was "intellectually such fun trying to make this work." They wanted to ensure that a senior citizen with an old furnace in their house would be protected. That's "why we brought in the climate credit," she says.

The British Columbia Low Income Climate Action Tax credit, or BCLICAT—a tongue-cramping acronym if there ever was one—gave back $100 to every adult and $30 for every child, per year, in each low-income household in the province. It was paid quarterly, added to the GST refund cheques that came like clockwork from the Canada Revenue Agency. It was to be increased annually at the rate of inflation. Taylor describes the BCLICAT as "a little bit of extra help to help [lower-income citizens] pay those extra costs." But while the help was real, it was essentially invisible to those who received the cheques.

Taylor was acutely aware of another fiscal constraint the new legislation would impose. The lower personal income tax rates, retroactive to January 1, 2008, wouldn't translate into dollars in the pockets of British Columbians until some time in 2009, once the Canada Revenue Agency had vetted the 2008 tax returns and sent out refund cheques. To circumvent any hardship that lag might have imposed, she led the development of a transitional "climate-action dividend," an advance, one-time payment of $100 for each adult and child, to be funded by forthcoming revenues from the carbon tax. She notes rhetorically that the income tax cut would take place "next year, but you are paying your gas costs as of July 1. So, how can I help you right now, until you get that tax cut?" Her answer was the dividend cheques, which went out at the end of June 2008, accompanied by a brochure that suggested "green" ways to spend the cash: add weather stripping, insulate crawl spaces, replace incandescent bulbs with more efficient compact fluorescent units.

That effort cost the Treasury some $440 million up front, before carbon tax revenues began to flow in. The idea was to help citizens transition to lower emissions. But some members of the public saw it less as altruism, and more as an attempt to buy political acceptance for the tax. Looking back, Taylor sees it differently. "It was a very important part of all the pieces we were trying to do to build consensus," she says.

Those pieces were many, a heady stew of variables, all of which had to be incorporated into the framework and the details of the tax structure. Andy Robinson and Glen Armstrong, two highly experienced bureaucrats within the Ministry of Finance, were the fiscal gurus who kept the effort on schedule throughout 2007. Robinson was assistant deputy minister of finance, while Armstrong was director of tax policy. Normally, they would have reported to the deputy minister in developing new tax policy initiatives and would brief the minister of finance, but the carbon tax was a bit different: Carole Taylor was directly involved in its construction. "We dealt with her directly quite a bit," says Robinson.[3] "She was totally comfortable dealing with tax and was not a minister who felt she had to have the deputy around very much on the tax stuff."

All of their work was done in secret by a team that numbered in single digits. Not even "other people in Treasury Board staff and finance that worked just down the hallway from us could hear or see any of the stuff we were preparing," says Robinson, noting that not even spouses were aware of what was being planned. None of this was unusual in the world of tax policy, he emphasizes. "It was just something you did. And you'd never talk to anyone about any tax changes coming up."

But two things made this effort different.

The first was speed: it was of the essence, and it brought with it pressure. The tax "was going to be designed very quickly and it was going to be implemented very quickly," says Armstrong. "A core group of people worked long hours to get it done," says Robinson, "but the team was very enthusiastic about the prospect of this policy, and so I think most people found it not a problem to work on it.

"They just sort of relished the opportunity," he continues. "It was a big policy change. It was clearly very innovative and forward-thinking. It hadn't really been done exactly like this anywhere else. And in my view, it definitely wouldn't have happened without Gordon Campbell."

The second was the revenue-neutrality provision. From the outset, says Robinson, "we were encouraged to make sure that [the tax] wouldn't make any money." Armstrong concurs: "The marching orders always were: this can't be a money-maker, so the offsetting tax cuts have to be greater than the [anticipated] revenue number each year." Revenue neutrality "made it more real for people," says Robinson. "Basically, it was a matter of choice," adds Armstrong. "You could choose to reduce your carbon emissions, but you're going to get this tax cut anyway." It was the Gordon Campbell philosophy, writ large.

Now both retired, the former civil servants look back on those pressure-filled days with a sense of great pride. "It was a highlight of my career," says Robinson. "You don't very often get a high-level London School of Economics professor saying [that] something you had a hand in producing is a template for the world."

Carole Taylor, too, shares pride in bringing down taxes for both people and businesses, but the math showed that the benefits went "two-thirds to the people, and one-third to business, because we all know business passes costs on," says Taylor. "So they weren't going to get hit straight in the head the way individuals would." Her finance ministry officials "ran different cars, different trucks, different everything, trying to figure out what the real costs would be, how much [individuals would] get back with the income tax, and it would be at least covered. [People] could decide to make some personal changes that would give them some extra money, or [make] no changes and they would be flush." Taylor looks back now and says, "I love the model. I see everybody fiddling, but I think it's the best in the world."

She'd get no argument from Paul Ekins.

For BC's citizens and companies, there can be no argument that the carbon tax represented a tax shift and not a scheme to generate new net revenue. In Gordon Campbell's mind, that shift firmly cast the spotlight on personal choice while encouraging conservation—individuals or companies that reduced fossil-fuel consumption benefited from a lower overall tax burden, since income taxes were reduced regardless. The revenue-neutral design was a direct manifestation of the Campbell philosophy—big car or small car: you choose.

Revenue neutrality offered two other big pluses, one politically smart, the other economically intelligent.

Politics first. Martyn Brown describes the rationale for the revenue redistribution as being "innately political," noting that Carole Taylor "was always looking for ways to make things more politically palatable." And in terms of anticipated pushback from the general population, Brown notes that the revenue-neutrality requirement was "a saw-off that you could point to." Moreover, it hobbled the left-wing NDP, which prided itself on its progressive environmental policies. How could the New Democrats complain that the revenue-neutral carbon tax was just another grab by government?

The fiscal effects of lowering corporate and small-business taxes were expected to minimize adverse impacts on BC's economy. The mining, forest and natural gas sectors were widely expected to push back, but Brown points out the political reality of the day: the resource-development community had no grounds on which to do so. Taxes were low, the regulatory burden was very low, and Brown argues that the government had already delivered what the business community wanted: "It wasn't in a position to say, 'We didn't get anything from you guys.'"

Some predicted the sky would soon fall once the tax was announced. The Canadian Taxpayers Federation—a group known for its strident low-tax positions—was quick off the mark. The day after Carole Taylor stood in the House and made history, the federation's BC director, Maureen Bader, complained to CBC News: "The government was elected on a tax- and regulation-reduction mandate. But this so-called revenue-neutral carbon tax will be anything but neutral for individuals, businesses and industries. It will create hardship for families, as soccer moms are unlikely to start walking."

Bader failed to point out that the amount of income tax paid by any soccer mom or soccer dad was to be decreased, retroactively, to January 1, 2008. Soccer moms or hockey dads, if they drove less, carpooled more, or switched to a more fuel-efficient vehicle, would soon have *extra* cash in their pockets, thanks to higher net pay. Bader also failed to grasp that BC-registered corporations, both small and large, would be paying significantly less tax a mere four-and-a-half months later. The simple substitution of one tax for another—Gordon Campbell's tax shift—should not have been a difficult concept to grasp, but apparently it was.

The Official Opposition wasn't much better. NDP finance critic Bruce Ralston said the new tax "will hit consumers and average families the hardest, as large industrial polluters get a pass and a handout."[4] He was wrong. The income tax rate for "average families" was cut *before* the carbon tax took effect, and large industrial polluters—whose corporate tax burden was about to decline—would soon pay the tax on every tonne of CO_2 they emitted from fuels they burned. There was to be no pass, and there were to be no handouts. Looking back, Campbell is clear on that issue: "We were told by virtually the entire business community that they wanted no free riders."

Others—the usual suspects, some might say—got it right. Ian Bruce, a climate-change expert at the David Suzuki Foundation, praised Carole Taylor's budget, saying that carbon taxes are "one of the most powerful incentives we have to encourage companies and households to actually pollute less". Moreover, he implicitly understood the structure of the new tax *and* the importance of choice, pointing out that it "allows all of us to be part of the solution."

What is surprising is that so many others let misinformation, ideology or cheap oppositional politics suggest the tax would somehow be punitive, conveniently ignoring both global warming and the obvious benefits of tax shifting. On that latter front—failing to trumpet the benefits vigorously—there is blame to go around, and some of it can be assigned to Campbell's government. In its zeal to get the tax in place and to start blunting the exponential rise in greenhouse gas emissions, the government did not communicate strongly enough the critically important revenue-neutrality provision. It duly made the announcements, but those weren't heard by large chunks of the population.

Stories abound of individuals not understanding how a tax could be revenue-neutral, nor understanding that the government's greedy hand was *not* reaching yet again into the pockets of hard-working British Columbians. Many complainants did not understand that the personal income tax rate was reduced retroactively to January 1, 2008. That move gave citizens a head start on the tax. Others didn't know that if they had very little income, they'd receive a cheque every three months to compensate for the tax on their fossil-fuel consumption.

The tax shift was simply Greek to others, many of whom might have been expected to know better. One grousing resident of North Vancouver, an engineer no less, even made the pages of the *New York Times* in 2011, saying: "I've already insulated my house to be energy efficient. I already turn down my thermostat. Why should I have to pay $20 on my natural gas bill for something that is doing nothing for me?"[5] It didn't register that his income tax rate had dropped significantly more than three years earlier. His slightly higher natural gas bill was paying for the lower tax on his earnings. But was that shift well-communicated by the government? No.

Some farmers also complained. One highly intelligent and innovative former orchardist in the Okanagan, who invested heavily to convert from apples to grapes and built an award-winning winery, put it this way in 2014: "Every time I fill up my pickup, that goddam carbon tax costs me another four bucks." It was pointed out to him that his personal income tax had declined, the corporate income tax paid by his winery had declined, and his wife's income tax had declined. He replied, "But I don't see any of that. All I see is the four bucks more that it's costing me to fill up my truck." For him, revenue neutrality was close to myth.

Some quarters of the private sector had another concern, gnashing their teeth on the perception that the price on carbon would put BC industry at a competitive disadvantage. Carole Taylor was very aware of that concern and confronted it head-on, noting that while the tax would at some level influence competitiveness, lower corporate tax rates would offset that. "The very first speech I gave after delivering the budget was to the Business Council of BC," she remembers. "I talked it through and explained what taxes we were lowering, the certainty they would get, the fact that they would be able to say within three years we have the lowest corporate tax in Canada—not immediately, but in three years. All of a sudden, you could see the wheels turning." Jock Finlayson, the chief economist at the Business Council, was there. Taylor recalls wryly that "Jock had not been keen on the carbon tax," but at the end of her speech, "he came up and said, 'Well, I have to say, it's a good model.' And I thought, 'Success!'" The council was supportive, said Taylor, even though "they didn't love it."

Jock Finlayson's early conclusion that the tax was a good model came with a caveat. Since those days of February 2008, he'd penned frequent opinion pieces on the competitiveness issue. In 2011, he complained to the CBC that other jurisdictions were not following BC's lead.[6] "We are dancing alone," he said. Four years later, he told the *New York Times* that the Business Council's position had changed, reiterating their initial skepticism but adding that now, "within the business community there is a sizable constituency saying this is okay."[7] But by "okay," he didn't mean BC should charge ahead at full steam and raise the tax beyond $30 a tonne, where it sat at the time. "We're already head and shoulders above anybody else in North America," said Finlayson.[8] "We don't think it will actually be good for the economy to further widen the gap between carbon prices in BC and other jurisdictions."

Finlayson's concerns raise an important question: Did BC's economy suffer during the escalation phase of the carbon tax and after, when other jurisdictions were not following suit? There are two additional and equally important questions: Did our fossil-fuel consumption decline, and if it did, can the decrease be attributed to the impact of the tax? The answers to all three questions are simple: no, yes and probably.

Let's look at emissions first. The carbon tax was imposed at the midpoint of 2008 and increased on July 1 each year, with the last scheduled increment being applied on July 1, 2012. By June 30, 2013, after five full years of the tax being in place, per capita consumption of fossil fuels in British Columbia had fallen by just over 16 per cent in absolute terms (see Table 2), and by just over 19 per cent relative to the rest of Canada (RoC). The contrast between BC and the RoC is stark. In terms of changes in per capita fuel use, BC outperformed the RoC every year since the carbon tax came in. That performance apparently did not impress Campbell's successor, BC Liberal Premier Christy Clark, who announced in the spring of 2013 that the tax would remain frozen at $30 per tonne until 2018.

Some critics have suggested that during the four years in which the tax escalated in BC, fossil fuel consumption fell more than in the RoC because the global recession of 2008–09 hit the province harder. British Columbia was indeed walloped by the recession: provincial

TABLE 2. Per cent changes in per capita consumption of petroleum products (gasoline, diesel, fuel oil and propane) in British Columbia and the rest of Canada between July 1, 2007, and June 30, 2013. With the exception of the final column, the data are reported as year-to-year differences between twelve-month July-through-June averages. *Data from Statistics Canada, as reported in Pedersen and Elgie (2015).*

	2007/08 –2008/09	2008/09 –2009/10	2009/10 –2010/11	2010/11 –2011/12	2011/12 –2012/13	2007/08 –2012/13
British Columbia	-3.5	-6.7	-1.3	-6.2	0.2	-16.1
Rest of Canada	-2.5	-1.5	6.8	-1.7	1.7	3.0
Difference	-1.0	-5.2	-8.1	-4.5	-1.5	-19.1

GDP fell from an average annual growth rate of 3.49 per cent over the 2002–07 period to 0.97 per cent in 2008 and minus 2.58 per cent in 2009, before recovering in 2010 to plus 3.3 per cent.[9] Undoubtedly, the decline in both industrial production and demand for the province's raw materials had a negative impact on fuel use during the depth of the recession.

But other provinces went through a similar downturn. Comparing their fuel consumption data with those in BC helps to normalize out the impact of the recession, while casting the spotlight on the carbon tax's possible influence. The difference data in Table 2 compare the change in annual per capita consumption of fossil fuel products in British Columbia since 2007–08 with the same variable aggregated over the RoC. In two of the three post-recession years, per capita fuel consumption increased in the RoC while it declined in BC. Over the course of the four years of the escalating carbon tax, per capita fossil fuel use in BC fell by just over 19 per cent relative to the rest of the country. The contrast is dramatic. The recession cannot be blamed for the relative decline in BC's fuel consumption immediately following the introduction of the tax.

A key question remains. Can the clear decline in fossil fuel consumption in BC then be unequivocally attributed to the carbon tax? The simple answer is no, not unequivocally. Writing on behalf of the Business Council of BC in 2013, Jock Finlayson posed a trio of other potential influences, *exclusive of the carbon tax*, that might have driven down aggregate fossil fuel consumption in the province: a) the collapse of US housing markets in 2008 and a consequent decline in fuel demand by the industrially dominant BC forest industry; b) improved rapid transit coincident with increases in parking costs in the Greater Vancouver region; and c) a big jump in the number of British Columbians crossing the border into Washington State to fill their cars' gas tanks.[10]

Analysis of available data shows Finlayson had some fair points. First, heavy-duty diesel trucks are endemic in the forest industry. But total emissions from that vehicle type in BC *rose* over the course of the rising tax, from 4.9 million tonnes (Mt) in 2009 to 5.2 Mt in 2011, 5.7 Mt in 2012 and 6.4 Mt in 2013—no sign of a collapse in fuel consumption there.[11] And while lumber exports[12] from BC to the US declined by nearly half, from 23 million to 13.6 million cubic metres between 2007 and 2013—thanks to the implosion and slow recovery of the housing market south of the border—exports to mainland China skyrocketed more than twelvefold between 2007 and 2013, from 0.6 to 7.9 million cubic metres. Neither export levels nor aggregate fuel-consumption data point to the forest industry as a lead player in reducing fuel demand—and thus provincial emissions.

Maybe the lower emissions instead reflected a collapse in the pulp and paper industry, which has a high demand for both heat and electricity, some of which is supplied by natural gas consumption? BC's greenhouse-gas inventory data do not specifically break out the pulp industry, but it is a big player in the "Manufacturing Industry, Stationary Combustion" category. Emissions from that industrial cohort rose slightly, from 4.1 Mt in 2008 to 4.3 Mt in 2012 and 2013. Here again, the data offer no support for the suggestion that changes in the forest industry can explain the per capita decline in fuel consumption, relative to the rest of Canada, observed in the province between 2008 and 2013.

Perhaps rapid transit improvements in the Greater Vancouver region were a contributor? Certainly, the opening of the fifteen-kilometre electric Canada Line in August 2009 made commuting into Vancouver from suburbs to the south easier, and ridership exceeded expectations. But was it the carbon tax that encouraged commuters to leave their cars at home and use transit, or could it have been something else—simple convenience, for example?

Jim Johnson says the carbon tax had an impact on passenger vehicle use, but barely. Johnson is a Victoria-based consultant and former employee with BC Statistics. In 2015 he used data up to and including 2014 to address two simple questions: Did the carbon tax reduce vehicular fuel usage in Metro Vancouver, and if so, by how much did fuel consumption decline? His analysis considered multiple factors that could have influenced fuel use, including types of vehicles driven over the years while the tax was increasing, access to transit, historical transit fares, population density, age and sex of drivers, the escalating tax itself and so on.[13] All such variables, when appropriately weighted and considered in aggregate, can affect fuel use.

When Johnson ran his model *with and without* the carbon tax, with all other variables remaining unchanged, he calculated that the tax reduced annual greenhouse gas emissions in the Metro Vancouver area by at most 1 per cent. He drew a simple conclusion: "The carbon tax does work—just."

Nic Rivers and Brandon Schaufele came to a stronger conclusion. Economists at the University of Ottawa and Western University, respectively, they undertook a rigorous and detailed economic analysis of the tax's impact on gasoline consumption in British Columbia as a whole, not just across the Greater Vancouver area.[14] Their statistical modelling ruled out transit improvements as a significant factor in reducing fossil fuel consumption. Moreover, they showed that a five-cent-per-litre carbon tax had a fourfold greater impact on reducing passenger-vehicle gasoline consumption than did an equivalent five-cent jump in the market price.

That behavioural response is very clear in their comprehensive analysis, but it is also difficult to explain. Rivers and Schaufele suggest it may be embedded in psychological economics, what they call social norms. When everyone is seen to be paying the same carbon tax, environmentally conscious drivers know they are contributing to the public good when they drive less. They also know that their responsible gesture cannot be subverted by free riders who take advantage of lower congestion on the roads by driving more, because those less responsible drivers are also paying the tax—they will pay more, while the responsible drivers both pay less *and* contribute to the public good. Rivers and Schaufele thus argue that the design of the tax encourages prosocial behaviour, which yields a more dramatic decline in emissions than an equivalent market-price bump.

Jock Finlayson offered another possibility in 2013. He suggested that the observed decline in provincial fuel sales might in part be due to British Columbians filling up their cars just across the border with cheaper American gasoline, the implication being that the carbon tax wasn't as important as many were suggesting. But were cross-border fill-ups due specifically to tax-avoiding refugees, or could they instead reflect the very high value of the Canadian dollar from

2009–13, coupled with the lower cost of most consumer goods in the United States? If you live in Vancouver or Surrey or Richmond and the loonie is near par with the greenback, why not make the three-hour round trip to Bellingham, Washington, to pick up that new TV for two hundred bucks less and, while there, fill up at an Exxon station? Many did just that.

But as it turns out, the thousands who ventured south and filled up had a trivial impact on provincial fuel consumption—and thus emissions. Two statistically savvy researchers, geophysicist Andy Skuce in British Columbia and Yoram Bauman, a PhD economist then in Seattle, took a hard look at the data.[15] Both concluded, independently, that cross-border fill-ups could account for only a 1 to 2 per cent drop in fuel purchases in BC, a small fraction of the ~16 per cent overall per capita decline witnessed since 2008 (Table 2). Their work was reinforced by another rigorous and academically dense analysis published four years later by the University of Manitoba's Chad Lawley and a PhD student, Vincent Thivierge.[16] Those authors conservatively concluded that "a five-cent-per-litre carbon tax reduced gasoline consumption by 5 per cent," adding: "Our results clearly demonstrate that households respond to a price on carbon" and that "the BC carbon tax operated as intended."

There's one more set of information at play here: there was a decline in use in the carbon tax years of *every* type of fuel covered by BC's tax policy relative to the rest of Canada (Figure 1). Petroleum coke—and, with few exceptions, natural gas—aren't utilized to power cars heading south to buy televisions in Bellingham, and so declines in their per capita usage cannot be attributed to cross-border shopping. So Finlayson and the Business Council of BC were right to ask the questions they did in 2013, but they were wrong to imply that the carbon tax was ineffective during the mid-2008 to mid-2013 window. Hard data tell us the opposite.

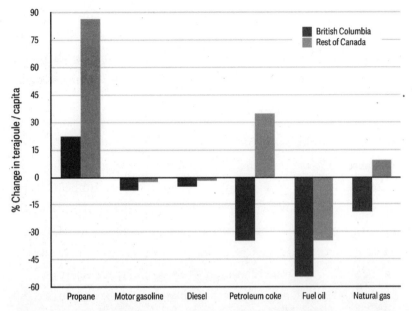

FIGURE 1. Percentage change in sales of specific fuels in British Columbia, reported as terajoules per capita over the six years beginning July 1, 2007. Note that the unusually steep decline in sales of fuel oil is attributed to refineries in BC switching production from "heavy fuel oil" to ultra-low sulphur diesel in response to new marine shipping and domestic transportation regulations on sulphur content of fuels. *Figure originally drafted by Sustainable Prosperity at the University of Ottawa, and is reproduced from Pedersen and Elgie (2015).*[17]

Over the years in which pundits were debating, and during which the data showing declining relative consumption was accumulating, British Columbia's media dutifully reported tax-driven increases in the price of fuel. But they often didn't get the context right.

In a stroke of coincidental but truly bad timing, the imposition of the first increment of the tax on July 1, 2008, coincided with a then world-record price for oil, which peaked at about US$132 per barrel at almost exactly the same time. The carbon tax pushed the price of a litre of regular gasoline in Victoria from $1.469 at one minute before midnight on June 30 to $1.494 two minutes later on July 1. Six months earlier, it had been just over a dollar a litre. Ninety-five percent of the increase since that time reflected the stunning rise in the world oil price.

That historical context didn't faze the *Victoria Times-Colonist*, however. Its bold headline on the front page of the July 2, 2008, edition read: "Gas prices push $1.50 as carbon tax kicks in." The implication was that the government was hammering Joe Public. Gordon Campbell was acutely aware of the bad timing. "If we had not had the highest gas prices that the world had ever seen on July 1, 2008, I think the carbon tax would have just gone through without any trouble at all," he said. "I mean there wouldn't have been any political flak."

Year after year, on or about the anniversary of the carbon tax, the media pressed the issue.

On July 1, 2010, the CBC posted "BC carbon tax jumps more than one cent."[18] That headline was, of course, wrong. The tax rose by $5 per tonne of carbon dioxide emitted, not a penny, but that $5 bump translates to a little over one cent a litre on gasoline. On July 1, 2011, the Canadian Press echoed the theme with a story entitled "Carbon tax bumps up BC fuels prices."[19] Exactly one year later, Canada's self-proclaimed national newspaper, *The Globe and Mail*, added yet another reinforcing headline: "BC to raise carbon tax, price of gasoline on July 1."[20] Over the course of five years, the British Columbia public repeatedly heard at the end of June or beginning of July that the price of fuels went up thanks to the carbon tax. Although it was seen by many as a negative reminder—who wants to hear that a tax is rising again?—the repeated use of the word *up* appears to have sunk in. It may well have been instrumental in shifting decisions at the individual level that translated into lower fossil fuel consumption.

For some reason, market-based jumps in price don't attract the same interest from the media, maybe because they don't carry political gravitas. As just one example, on October 30, 2017, the price of regular gasoline in Victoria jumped nine cents, instantly. There had been no warning, no foreshadowing. That nine-cent jump was two cents per litre more than the *full five-year increase* imposed by BC's carbon tax between 2008 and 2013—and yet the *Times-Colonist* printed not a word about it.

But we still cannot attribute reductions in fuel consumption unequivocally to the tax. Other possible influences remain. Maybe British Columbians are just more naturally committed to curbing global warming. After all, Greenpeace began in Vancouver in 1971, and the David Suzuki Foundation is an iconic institution in the province. Maybe BC's declining fossil fuel consumption relative to the rest of Canada during the years of the escalating carbon tax was simply the continuation of a pre-existing trend. If naturally more responsible environmental behaviour was key, it would suggest that the carbon tax was less effective than argued above, or maybe even ineffective.

To test this, sales data for all fuels subject to the tax are considered back to 2000 (Figure 2). British Columbians did use less fossil fuel per capita relative to the rest of Canada prior to July 1, 2008, in part because BC enjoys an abundance of carbon-free hydropower. But as Figure 2 illustrates, per capita fuel use in BC was actually rising slightly (by ~1 per cent per year) compared with the rest of Canada between 2000 and 2008. Following imposition of the carbon tax, fuel use in BC fell by ~4 per cent per year compared with the rest of Canada. This before-and-after shift in the data is clear, and argues against the possibility that the decline was simply the continuation of a pre-existing trend. The carbon tax appears to have trumped even BC's historically green inclinations.

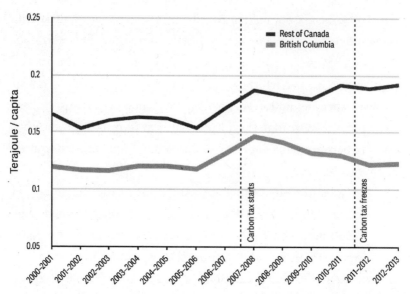

FIGURE 2. Aggregate annual per capita fuel sales in BC (light grey) and the rest of Canada (black) from July 1, 2000, to June 30, 2013. The vertical dashed line to the left indicates the date of imposition of the carbon tax, while the vertical dashed line on the right marks the beginning of the tax-freeze plateau (the tax was fixed at $30 per tonne CO_2e on July 1, 2012, a level that remained until 2018). *Figure produced by Sustainable Prosperity, University of Ottawa, and reproduced from Pedersen and Elgie (2015).*[21]

To this point, BC's carbon tax story is very positive. But here comes the kicker: *From its inception in 2008 to the point it was frozen in 2013, there was no indication that BC's tax had any overall negative economic impact.* Indeed, the gross domestic product of the province kept pace with, or slightly exceeded, that of Canada as a whole over the entire five years of the increasing price on carbon emissions (Figure 3). And that period included the downturn in demand for BC lumber when the sluggish US housing market began to collapse in 2008.

The carbon tax is, of course, only a tiny component of BC's overall economy, so one would not expect it to have a significant *direct* effect

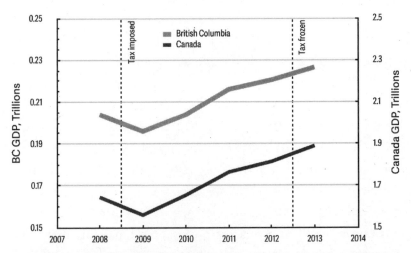

FIGURE 3. Annually averaged gross domestic product in BC and Canada, 2008 to 2013. Note the scales of the left and right axes differ by a factor of ten. *Data from Statistics Canada, May 2024.*

on its own. That is particularly true given that the revenue-neutrality requirement effectively recycles cash back into the economy, reinforcing in-province demand. Moreover, the lack of negative economic impact extends to broad sectoral levels. As one example, economists Brandon Schaufele and Nic Rivers, both then at the University of Ottawa, searched in 2014 for a discernible impact of the carbon tax on the export of agricultural products from BC.[22] They found none.

Energy-intensive niche sectors in BC, however, may well be more vulnerable to the effects of the tax. The greenhouse vegetable industry, for example, has traditionally used high volumes of natural gas to heat what is now over five hundred hectares of glass-enclosed land, most of it in the Fraser Delta in BC's southwest corner. The industry was vociferous several years ago, claiming that its cost of production was skyrocketing because of the carbon tax. In response, Carole Taylor's initial insistence that there be no exceptions was tossed aside, but for good reason: it isn't just heat that plants in greenhouses require for growth. They also need an abundant supply of clean ("pure food-grade") carbon dioxide to support optimum photosynthesis. And how was that being produced in BC? By burning natural gas. Switching to renewables like wood pellets or sawdust was not a good option for an industry that *needed* carbon dioxide as an input.

When the industry made that argument to the government in 2010, an exception was granted under the Carbon Tax Act. In 2012, the province announced that greenhouse growers would get a one-time, one-year carbon tax refund. After 2013 a different approach was taken: qualified producers could apply for grants that reimbursed 80 per cent of their carbon tax costs. That 80 per cent figure is important: by continuing to pay one-fifth of their carbon tax, the greenhouse operators had an ongoing incentive to increase fuel-use efficiency and reduce their use of fossil sources in favour of, for example, landfill-generated methane.

But the question remains: Is the greenhouse industry disadvantaged by the carbon tax? To some extent, yes, given that it must continue to combust natural gas to produce carbon dioxide for its operations. It is nevertheless holding its own, and it has been growing. Statistics Canada reports that greenhouse tomato production in BC increased from 43 million kilograms in 2007, before the tax took effect, to 63 million kilos in 2013, a 32 per cent increase in production that represented a 31 per cent increase in farm-gate value.[23] No other province increased its rate of hothouse tomato production as fast as BC during the 2007 to 2013 period.

But Linda Delli Santi, executive director of the BC Greenhouse Growers' Association, pointed out what remains an incontrovertible concern: "Our competitors don't have a carbon tax."[24] That comment would register well with Jock Finlayson. Until the rest of the world follows BC's lead, competitiveness will always be in the foreground.

BC's cement industry, too, is acutely aware of the internationally competitive environment in which it sells its product. Just like greenhouse vegetable growers, it is an intensive consumer of fossil fuels, largely because the manufacture of Portland cement requires large volumes of limestone (calcium carbonate) to be heated in enormous rotating cylindrical kilns, along with clay minerals or pulverized shale. The kilns are impressive behemoths, up to 180 metres long. Temperatures within them reach as high as 1,470° C, a quarter the temperature of the surface of sun.

It takes a lot of fuel to run a large, mineral-laden rotating cylinder twice the length of a Canadian football field, and at very high temperatures. If that fuel is coal or petroleum coke or shredded tires or natural gas, it brings with it a large carbon tax bill. The industry in British Columbia, which produced almost a fifth of Canada's cement output in 2008, wasted no time in complaining to the government that the tax threatened its competitiveness. "BC's unique carbon tax has hurt the cement industry's traditional ability to export cement powder into the Pacific Northwest," it wrote in a letter to the premier in 2011.[25]

But it wasn't just the carbon tax that was hampering cement sales from BC. China had been expanding its cement manufacturing capacity at an astonishing rate over the previous two decades. By 2016 it was producing thirty times as much cement annually as the entire United States, and it was cheap. Canada, the United States, South Africa and India, among other countries, were being flooded with low-cost Chinese cement. The Cement Association of Canada tried to spin the tax as somehow being responsible for the flood, writing: "Since the introduction of the carbon tax in BC in 2008, imports of foreign-made cement to the province have gradually climbed from less than 5 per cent in 2008 to a peak of over 40 per cent as the tax progressively increased to $30 per tonne."

Correlation is not causation, and the association's analysis ignored the fact that Chinese cement was essentially being dumped around the world. Imports went up because that foreign cement was dirt cheap, and that unavoidably and negatively influenced demand for BC-made cement. And there were other influences on demand for cement powder. New residential and commercial construction in the US—almost the sole export market for BC-made cement—declined markedly in the aftermath of the sub-prime mortgage debacle of 2008–09. On top of that, the appreciation of the Canadian dollar from early 2009 made our product even more expensive. Demand for Canadian cement south of the border plummeted as American buyers looked elsewhere, and China happily accommodated.

Nevertheless, the Cement Association had a point: BC's carbon tax did progressively increase the cost of manufacturing. That made it harder for cement producers in the province to compete with foreign suppliers, particularly those dumping product in Canada. The president of the Cement Association, Michael McSweeney, put it this way in 2011: "Imported cement is not subject to the BC carbon tax. Foreign cement powder comes into BC tax-free." There's no spin in that comment. It again reinforces Jock Finlayson's concern that dancing alone, at some level, can be damaging. But hinting that the carbon tax was a principal cause for cement kilns to be running at less than full capacity fails to recognize that the economics of cement production and export are not just a one-piece jigsaw puzzle.

With very few exceptions, British Columbia's now permanent carbon-emissions taxation program did not produce economic harm. There is irony in that statement, however. By focusing on the potential for harm rather than on benefits, it's possible to undersell the positive: the carbon tax was almost certainly a significant factor in stimulating development of the province's cleantech sector.

In 2020, KPMG reported continuing dramatic growth in the cleantech industry, a trend that had begun over decade earlier.[26] Jonathan Rhone, then chair of the BC Cleantech CEO Alliance, notes that the province had "emerged as a globally recognized centre of cleantech innovation."[27] Rhone recognizes that "the carbon tax, coupled with lowered corporate income taxes, was a clear contributor to that progress." Rhone adds that Gordon Campbell "inspired a lot of people in our sector to say, 'If I'm going to build a clean technology company or climate tech company, I'm going to do it in British Columbia.'"

The revenue neutrality was "a critical selling feature," Rhone argues, given that "we're all allergic to corporate taxes." As a leader in the Vancouver business community, he played a seminal role in explaining the carbon tax to the resource sector, telling colleagues, "Look, this does not have to increase your taxes. You're incentivized to reduce your carbon footprint. You can actually reduce your taxes." Offering that view made a difference particularly in the mining and forest industries, Rhone suggests, "because the carbon tax was revenue-neutral. That's the genius."

British Columbia's carbon tax was also a factor on the international stage. David Helliwell, co-founder of Pulse Energy, a leading software-based energy management company in Vancouver, saw it first-hand.[28] "From 2010 to 2013, being a cleantech company from one of the world's most progressive carbon-tax regulatory jurisdictions gave us international street cred by association," Helliwell says. He once gave a talk in London, organized by the *Economist* magazine. During his talk, he referred to BC's policies, and noted how they had not slowed growth: "I got a round of applause in the middle of my talk, which doesn't happen to me very often."

In 2013, Pulse Energy signed a major contract with British Gas. The deal was one of the largest ever done in the world of energy-management software, and it was notable for another reason: it allowed a former premier to help harvest a crop from seeds he'd sown five years earlier, because by then Gordon Campbell had been appointed high commissioner to the UK. He "was personally involved in helping to get British Gas executives on-board with the project," says Helliwell.

Gordon Campbell fought the 2009 provincial election in the spring of that year, about fourteen months after Carole Taylor's budget speech in the House and ten months after the carbon tax had taken effect. He and his party were re-elected to a third term. Looking back at the campaign, Campbell is matter-of-fact: "We won in 2009 because of our position on climate. It added to our other positions," he says. "Three or 4 per cent of the vote in British Columbia makes the difference between being a majority government or not, and I'm sure we got 3 or 4 per cent from people who really didn't care much about politics but they did care about the environment."

In his mind, the 2009 election was a triple win. "We won on the economy, we won on the environment, and we won the election," he says. "If you go back and look at some of the discussions in 2009, there were people saying, 'We have to support the BC Liberals ... because if we don't, no one will ever try a carbon tax again.'"

The experience in British Columbia shows that a carefully thought-through tax shift that penalizes pollution while rewarding conservation and improving environmental outcomes can be a winner. Paul Ekins could see that immediately when he visited BC in 2010. Two years later, he wrote to the new premier, Christy Clark, when the impact was being reviewed. In his letter he described BC's carbon tax as "among the best designed measures of its kind in the world." He was right.

But during the years when carbon-tax proponents were celebrating the success of BC's model, opportunistic politicians and global-warming deniers were scuttling attempts elsewhere to put a price on emissions. Nowhere was that truer than in Australia. The Australian experience with carbon taxation is black to British Columbia's white, chalk to its cheese. It is a sordid tale of brutish politics, public lies, knives thrust into political backs, and an intellectually corrupt media. What follows is the story of how Australia almost emulated British Columbia's success before the hard right flung climate policy off the edge of a cliff.

CHAPTER 4

Down Under I: The Hawke-Keating-Howard Years and Australian Climate Action Dithering

For over a century, British Columbia and Australia have shared many physical and social parallels. Both economies were initially sparked by gold rushes, and both have traditionally been rooted in resource exploitation. Both jurisdictions have seen rapid population growth. Both have benefited from immigration, and both have been socially progressive. Both have also been hit by the severe environmental vagaries of global warming: BC lost much of its pine forests in the first decade of this century; Australia suffered through "the drought of the millennium" at the same time, experienced unprecedented wildfires and water shortages, and is now seeing the survival of its iconic Great Barrier Reef be grievously threatened.

Environmental consciousness in Australia truly awakened in the 1980s, as it did in British Columbia, driven in part in both jurisdictions by growing concern about old-growth logging. That issue was almost a daily topic of discussion around dinner tables in BC then, as it was in Tasmania on Australia's temperate southern periphery and in Queensland on its tropical north coast; Australia had protests in both states. In BC, the "war in the woods" at Clayoquot Sound on Vancouver Island in August 1993 culminated in the biggest mass arrests in Canadian history at that time. In a twist of historical coincidence, Australia had a presence at Clayoquot Sound that summer, which had an impact on policy Down Under some fifteen years later: the iconic rock group Midnight Oil played a concert at the Clayoquot protest camp. Lead singer Peter Garrett—who in 2007 was appointed Australia's environment minister by Prime Minister Kevin Rudd—was moved by the old-growth clear-cuts he witnessed nearby and declared: "This is no way to look after the land."[1]

Forest management issues stimulated an environmental conscience in both countries, and pushed governments to respond with smarter policies. But another and much broader concern—one with far greater consequences for human society—was brewing in the background: global warming. Like magma beneath a volcano, that issue was silently building to a point where it would erupt onto the policy stage in both Australia and BC. Although there are many parallels between the two jurisdictions, the eruptions fundamentally differed. BC confronted the climate challenge head-on with a flood of legislation over an eighteen-month period. Australia's actions were marked by false starts and policy ephemera. That behaviour has persisted for three decades, to the present day.

The phrase *global warming* is actually relatively young. It is thought to have first appeared in the scientific literature in August 1975, when renowned American oceanographer Wallace Broecker published a paper with a question in the title: "Are we on the brink of a pronounced global warming?"[2] An affirmative answer would have immediate implications for climate policy worldwide. Nowhere was that truer than in Australia, the driest permanently inhabited continent on the planet. Only Africa is hotter.

Scholars at the Australian Academy of Science, stimulated by Broecker's question, took up the challenge. Could Australia become even drier and even hotter? Seven months after Broecker's paper was published, the academy reported that whereas human activities are likely to contribute to warming *in time,* "there is no evidence that the world is now on the brink of a major climatic change."[3] In Australia in 1976, drier and hotter were still considerations for the future.

History hasn't recorded whether or not that conclusion led to a sigh of relief—but regardless, there is no record of global warming becoming a political concern until a decade or so later. Economist Ross Garnaut, author of the famed Garnaut Climate Change Review of 2008,[4] advised Prime Minister Bob Hawke's government on economic policy from 1983 to 1985, and said years later that global warming in the early 1980s "was not much on our radar, his or mine."[5]

Labor stalwart Bob Hawke had cut his political eye teeth as the long-serving head of the Australian Council of Trade Unions before winning a seat in the federal Parliament in 1980. He became party leader in early 1983 and led a landslide victory over Malcolm Fraser's Liberal Party in the national election that March.[6] As Ross Garnaut confirmed, Hawke was not particularly environmentally minded in those years. But by the late 1980s he was well into his eight-year tenure as prime minister, and the global-warming radar screens had been lighting up—both in public and, importantly, on the backbenches of Parliament.

In October 1987, federal Labor MP Robert Chynoweth delivered an impassioned speech in Parliament.[7] He began simply: "I wish to speak today on the environment," before going on to offer prescient observations. He recounted what scientists warned awaited Australia: changing rainfall patterns, increased temperatures, rising sea levels, changes in the character of cyclones, more common floods and droughts. He understood scale: "We are about to be overcome by a dilemma that pervades almost every aspect of our very existence." And he had an obvious appreciation for fiscal dimensions, asking, "Will the benefits of economic growth be cancelled out by climatic changes?" and "Who will pay for the cost of coastal destruction?"—a question that hinted at Sir Nicholas Stern's later characterization of global warming as "the greatest market failure the world has seen."[8] Chynoweth closed with three sentences, the second of which would describe far too much of Australian climate politics a quarter-century later, saying that the whole world is "the life-support system crafted by nature over billions of years. It is waiting to be redeemed by us, and not crisply toasted—a toasting brought about by ignorance, a lust for profits and a lifestyle that can easily be changed. We must act now."

By mid-1989, environmental issues were top-of-mind in Australia: not just global warming and ongoing deforestation, but ozone depletion. All were on the front pages on an almost daily basis.

Senator Graham Richardson, known in Australia as "the Fixer," was a key cabinet minister in Hawke's government then. A colourful politician described by journalist Marian Wilkinson as "a backroom political knee-capper" and "the most powerful right-wing headkicker in Canberra," Richardson was no friend of the growing environmental movement.[9] In 1986 he even branded environmentalists as "the lowest form of life on Earth."[10] But in the austral autumn of that year, he underwent a transformation, one that led him to become, in Wilkinson's words, "a Labor Party hero."

Bob Brown was behind that transformation. A general practitioner who sat in the Tasmanian Parliament, Brown served as a Green Independent. Years later, he went on to lead Australia's Green Party, but in April 1986 his concerns were more provincial. Tasmania's southern forests were being decimated by logging companies, including Australian Newsprint Mills, described by Wilkinson as being "one of the largest logging and pulp mill operations in the state." Richardson had toured their operations and came away impressed. Brown was leading the fight to preserve the forest.

Brown versus Richardson: chalk versus cheese, one might think. But in a eureka moment, Brown and his colleagues invited Richardson to tour southern Tasmania with them by helicopter, offering to show him what Australian Newsprint Mills did not. As Wilkinson recounts, Richardson witnessed great chunks of desecrated flattened forest littered with huge piles of wasted timber. Humans—not mountain pine beetles, like in British Columbia—had infested the former stands of mountain ash in Tasmania. Overnight, Richardson's position on forestry practices switched from being sympathetic to big corporations to appreciating that the destruction could not go on.

It wasn't sentimentality that drove that switch. It was common sense, coupled with political smarts. Richardson was well aware that "public opinion was shifting in favour of the environment."[11] He carried his thoughts back to Canberra and arranged for Bob Hawke to meet with Bob Brown, and a year later the Hawke government suspended rapacious logging operations in the southern Tasmanian forests. The Fixer had come through, at least in the eyes of the environmental movement.

The forests in Australia's southernmost state were only one front in the 1980s battle to preserve the biological heritage in the country. In the north, an even harsher fight was under way, centred on the tropical rainforests of northern Queensland. Small in area, they were being severely overcut under the policies of the long-serving (1968–87) conservative premier of Queensland, Joh Bjelke-Petersen, a fan of intensive old-growth logging.[12]

Nicknamed the "Hillbilly Dictator" by the media, Bjelke-Petersen fought Hawke's efforts to preserve at least a fraction of the Queensland tropical rainforests, but he lost in the end.[13] Richardson described the issue as a "political tidal wave only likely to get bigger" and lobbied Hawke to take action.[14] He did. In 1987 his government nominated the wet tropical forests of Queensland for listing as a World Heritage Site, thus guaranteeing their future preservation.

Meanwhile, destruction of another type was taking place in the Southern Hemisphere. Ozone was being lost from the thin air—the stratosphere—high above Antarctica. Stratospheric ozone is like a screen that absorbs solar ultraviolet (UV) radiation, reducing the ultraviolet flux that would otherwise make it down to the land and ocean surface at levels that would damage both flora and fauna. In the late 1970s it had been discovered that man-made chemicals were upsetting the stratospheric ozone balance. Chlorofluorocarbons, colloquially known as CFCs and widely used as propellants in aerosol cans or as refrigerants in cooling systems, were chemically reacting with ozone and consuming it. This was most prominent over the South Pole, owing to a peculiar geographic confluence of seasonal changes in sunlight levels and atmospheric circulation.

Australia isn't that close to Antarctica, but it's also not that far, and the growth of the "ozone hole" over Antarctica was enough to drive concern that more ultraviolet radiation might soon reach Australian beaches, playgrounds and backyards. In moderation, ultraviolet radiation can give one a nice tan, but too much increases the risk of skin cancer and can harm vision by stimulating cataract development. Those prospects are personal—not remote, like rising sea levels or declining river flows—and they clearly sharpened environmental consciousness in Australia in the 1980s. As a result, the country was one of the first to ratify the Montreal Protocol, the 1987 international treaty that established a mandatory schedule for phasing out ozone-depleting chemical compounds.

The Canadian connection here is one of which we can be proud: our country played an instrumental role in crafting the Montreal Protocol. Elizabeth May, who later became leader of Canada's Green

Party, wrote that Canadian lawyers chaired the group that drafted the text, and that the country's delegations "were champions" of the agreement.[15]

That wasn't the only Canadian climatic connection with Australia in those days. At the June 1988 conference in Toronto—"The Changing Atmosphere: Implications for Global Security"—some three hundred scientists from forty-six countries, including Australia, came together by invitation to discuss both the science of climate change and policy responses that would address it.

Canadian Prime Minister Brian Mulroney opened the meeting. Mulroney was an environmentally conscious conservative who, like Gordon Campbell but unlike many on the right wing today, understood that a true conservative has an obligation to conserve. Also playing a formative role at the conference was Stephen Lewis, an oratorical genius and democratic socialist who, despite his politics, had been appointed Canada's ambassador to the United Nations by Mulroney in 1984. Described as having brilliant persuasive powers, Lewis brokered the astonishing final declaration: "Humanity is conducting an unintended, uncontrolled, globally pervasive experiment, whose ultimate consequences could be second only to global nuclear war."[16]

That statement registered almost simultaneously four thousand kilometres away in the Vancouver mayor's office and sixteen thousand kilometres away in Australia. But the responses in the two jurisdictions differed. Gordon Campbell created the regional task force that recommended steps to curb emissions. In Canberra, Graham Richardson, the "right-wing headkicker"—tried to light a similar fire under Bob Hawke's government. Unlike Campbell, he was to fail.

By openly recognizing that "ultimate consequences" were the end point on a long greenhouse-gas emissions pathway, the attendees at the Toronto conference stepped into the policy domain. They advocated what came to be known as "the Toronto target," a global reduction of greenhouse gas emissions of 20 per cent from 1988 levels by 2005. In 1989, Richardson—publicly describing the greenhouse effect as "the greatest threat facing Australia and the world"—submitted the Toronto target for consideration by Hawke's cabinet.[17] But economic concerns dominated the subsequent discussion. In the end, and over Richardson's objections, the Toronto target was crushed by Australia's treasurer and future prime minister, Paul Keating.[18] The writer Maria Taylor, citing a financial review from November 1989, notes that Keating "convinced cabinet that Australia should instead promote itself as an energy-efficient industrial centre."[19] The term *energy efficient*, when Australia's industrial and electricity-generating sectors were built around coal, evidently meant something different than it does today.

Despite Keating's aversion to action, the environmental genie was then out of the bottle—and Bob Hawke knew it. In July 1989, he ran the environmental banner firmly up the flagpole when he delivered to the nation his "Statement on the Environment: Our Country, Our Future."[20] Hawke, not known for overstatement, described the sixty-six-page document as "the world's most comprehensive statement on the environment."[21]

The statement captured societal obligations that too many governments had been shunning. With a nod to a long-term view that embraced both intergenerational equity and international responsibility, Hawke wrote: "Australians have a responsibility to preserve the unique ecosystems of this continent and to play a part in maintaining the Earth's biological diversity." He firmly recognized that while the Australian constitution grants significant responsibilities to the states and territories, like Canada with its provinces, the Australian business community was stridently averse to the development of state-level environmental policies that would introduce complexity more than they would provide national-scale solutions. Hawke agreed: "Many of the environmental problems we face today do not respect state and territory boundaries, and cannot be resolved piecemeal. Increasingly the Australian community and investors are demanding national approaches to major environmental issues ... They do not want as many systems for dealing with these problems as there are states and territories."

He went so far as to suggest that should public awareness reach "a sufficiently high level ... the Government would consider proceeding with a referendum addressing the constitutional powers of the Commonwealth over the environment." In offering that thinly veiled threat, he made it clear that he wasn't about to let the Australian states dictate regional environmental policies that could constrain national action. Some Canadian premiers today would do well to heed those words.

Importantly, "Our Country, Our Future" included major discussion of climate change. Hawke was well aware of Australia's international obligations, writing: "The Government will take an active role in developing a new framework convention on climate change," and "Australia will continue to give the Intergovernmental Panel on Climate Change strong support." He went on: "Australia will play its part in reducing global greenhouse gas emissions, both domestically and internationally," and he reiterated the 1988 Toronto call for a reduction in carbon dioxide releases to the atmosphere.

But too often, words, like good intentions, evaporate under the intense spotlight of economic concerns, particularly when those are rooted in preservation of the status quo. Hawke's cabinet, in particular Paul Keating, was loath to support legislation that would drive carbon dioxide emissions down. Others were influencing national opinion in the aftermath of the Hawke statement. As Maria Taylor describes it, "industry lobbyists, free market economists, and trade bureaucrats were winning with an argument that the fossil fuel economy must stay as it is (being Australia's 'natural advantage') and that Australia should take no action until other countries did."[22] That refrain, to wait for others, might sound familiar to British Columbians in recent years, for it was commonly used by former premier Christy Clark as an excuse to avoid taking further climate action in BC.

By the austral spring of 1990, the potential policy pendulum in Hawke's administration was shifting away from pure environmental concerns. Ros Kelly, who had replaced Graham Richardson as minister in April 1990, adroitly captured the tone. Referring to economic development while respecting environmental concerns, she echoed Gordon Campbell, saying: "It is not a case of one of the other; it is the integration between the two."[23] That integration, perhaps better described as a marriage, was formally announced a month later when the Hawke government adopted the Toronto target, but only after Keating insisted that a "no-regrets" proviso be added: "The [emissions] reduction would not be at the expense of the economy."[24]

Despite "economy" being in the ascendency in those months, Australia remained keen to show international environmental leadership. Minister Kelly rose in Parliament in November to describe the position Australia had taken at the second World Climate Conference, held in Geneva just days earlier: "We also said it was not enough for our country to have a sensible position in relation to greenhouse gases; it was essential that our position be seen as part of a very strong stand by other industrialized nations."[25] Australia under Bob Hawke still saw itself as a key member of the club of countries pushing for a global agreement to limit greenhouse gas emissions.

It was not to last.

On December 19, 1991, Paul Keating became prime minister of Australia, winning a spill that marked his second challenge that year to Bob Hawke's leadership.[26] The global recession of the early 1990s was punishing Australia's economy, and that, coupled with Labor infighting, put Keating over the top in an MPs' internal vote. With Hawke's relegation to the backbenches, just a week before Christmas, Australia essentially declared a time-out from championing climate action. Labor firmly put economy before environment, conveniently forgetting Ros Kelly's words in the House sixteen months earlier: "It is not a case of one or the other."[27]

Keating's Labor government simply wasn't prepared to build on the actions of the Hawke-Richardson era. Other concerns dominated, among them reforming tax and family benefits programs, formally reconciling with Australia's Indigenous populations, establishing Indigenous rights, and putting in place a national pension scheme. "The momentum was lost when Paul Keating became Prime Minister in 1991; his government showed little commitment to climate change," wrote well-known Australian author and political authority Clive Hamilton in 2007.[28]

It was following that period of neglect that the Liberals won the 1996 federal election and John Howard became prime minister. A conservative in every sense, he held the office for eleven years. Howard's writings paint an appropriate picture of the extent to which he perceived the environment and climate change as priorities: he didn't. For the six years from 1989 to 1994, he wrote columns for the dominant newspaper chain in Australia owned by arch-conservative Rupert Murdoch. George Megalogenis, a seasoned veteran of the Canberra press gallery, covered politics through those years; in his book *The Longest Decade* he offered a top-ten snapshot of Howard's political convictions gleaned from his published newspaper prose. They included economic management, Aboriginal reconciliation, industrial relations reform, motherhood and six others. Among the collection there is no mention—zero—of the environment, global warming or climate change.

Guy Pearse, author of *High and Dry*—a comprehensive account of John Howard's climate record—is less generous, describing Howard's response as "a farce."[29] Pearse, a former Liberal insider, speechwriter and lobbyist, was raised in the 1970s in "a very special part of the world," the north Queensland coastal city of Townsville, known for its proximity to the Great Barrier Reef.

Neither he nor John Howard would recognize the reef today. Global warming has pushed temperatures beyond the breaking point for coral survival. Many reefs in the waters directly east of Townsville are dying. In 2016, two-thirds of the corals toward the north end of the Great Barrier system died. By the late austral summer of 2017, the picture had not changed: more reefs were succumbing to surface-water temperatures well beyond historical averages.

John Howard was warned of this eventuality by his own government scientists in the 1990s: the future of the Great Barrier Reef was a central concern of all climate-change projections then, as it is today.[30] But Howard dithered on emissions policy, choosing to do as little as possible. His government spent just $22 million per year on climate programs in its first two years.[31]

At the time, Guy Pearse was a graduate student soaking up a wide range of political perspectives in Boston at Harvard's Kennedy School. He records that the Australian Liberal Party in the mid-1990s "had little interest in environmental policy ... It was viewed as enemy territory, an issue that at best we needed to neutralise."[32] But his Kennedy School experience led him to question that position, as he came to realize that economic prosperity and environmental responsibility were not mutually exclusive, despite what many of his colleagues back home thought. Case-study research at the Harvard Business School in 1996 confirmed his new-found views: "More often than not, environmentally benign business practice is just as profitable as environmentally harmful business," he wrote. And in echoing

Gordon Campbell's philosophy, he added that it was "market mechanisms that could most effectively protect the environment."[33] That thinking was slowly seeping into some elements of the Australian Liberal Party. And there was one advantage to having environmental policy be a low priority: it had nowhere to go but up.

Robert Hill, a senior senator from South Australia first elected in 1980, was the man to push it there. John Howard had appointed him minister of the environment following the 1996 election. Hill was a savvy, small-L liberal, highly knowledgeable about his portfolio and concerned about climate change. He was also government leader in the Senate; he had power. And by early 1997 a new impetus was on the scene: planning was under way for the Kyoto Conference scheduled for December of that year, and Australia needed to get all its negotiating ducks lined up. Thanks to growing international momentum leading up to Kyoto, benign neglect was to be pushed aside in favour of doing some good.

That good was going to be hard to find. Clive Hamilton described what was happening in the background in the months before the conference: the fossil fuel industry was pushing for Australia's negotiating position to be toughened, going so far as to bring hardcore global-warming deniers from the US to a conference in Canberra in late August entitled "Countdown to Kyoto."[34] One aim was to "offer world leaders the tools to break with the Kyoto treaty." ExxonMobil, a major supporter then of the deniers, was one of the sponsors. That same month, at least seven senior government ministers flew to Japan to argue against common emissions-reductions targets for developed nations, instead lobbying for differentiated national reduction objectives. For Australia, the objective was entirely self-serving and rooted in three key considerations: a rapidly growing population; the primacy of coal-fired electricity production within the country; and its position as the world's largest exporter of coal. That combination sat behind the Howard government's concern that emissions reductions would be too hard-won. By sending ministers to Japan, Howard was working to soften up the global community so that it would be more receptive to Australia's negotiating demands four months later.

The Kyoto Conference in late 1997 was the third major international gathering held under the auspices of the United Nations Framework Convention on Climate Change (UNFCCC), which had been hammered out at the 1992 Earth Summit in Rio and formally ratified by 197 nations in 1994. Kyoto was to be a turning point in stemming the rising tide of greenhouse gases. As a highly developed country and one of the planet's highest per-capita emitters, Australia was expected to play a key role. After all, the country held a reputation, going back to Bob Hawke, as a nation that understood the climate challenge.

But Robert Hill went to the negotiations burdened with a holdover from the Keating government: Australia was not to accept any treaty provisions that would jeopardize the economy—period. Howard himself added another complication: choosing stealthily to underplay developed-country obligations, by publicly announcing a week before the conference that developing (read: poor) countries needed to be involved in any solution to global warming.[35] That was code for Australia's unwillingness to accept its share of the emissions legacy, and the associated responsibility to do something about it. It wanted to duck any role in climate leadership in Japan that December.

And duck Australia did. Robert Hill went to Kyoto not to encourage the development of an international agreement but to insist that Australia be given special treatment. Like a glutton demanding a bigger plate at the dinner table, he went armed with a cabinet directive that Australia be permitted to *increase* its emissions to 108 per cent of 1990 levels by 2012. Meeting that level would require that Australia be given full credit for reducing land-clearings rates in the country. At the eleventh hour, Hill demanded that a clause in the agreement be inserted to that effect. The clearing of land, largely for expansion of agricultural operations in Queensland, had peaked in 1990, something Hill knew very well. By setting 1990 as the baseline, which was the plan for Kyoto, the agreement's inclusion of land clearing would give Australia free room to expand its industry emissions in the coming years. It was sleight of hand, a game of chicken that Hill played well. Fearful of the protocol foundering on a rather vague concession to Australia, delegates from other countries acceded to Hill's demand. In an astonishing turn, the "Australia Clause" was added to the agreement at the very last minute.[36] The Kyoto Protocol was born.

Hill was greeted with a hero's welcome on his return to Australia, but elsewhere his Kyoto tactics were sharply resented. Hamilton recounts the European negotiators' response in particular: Australia "got away with it," said the European commissioner for the environment, the Danish politician Ritt Bjerregaard.[37] Peter Jorgensen, spokesman for the European Union, described the increase as "wrong and immoral." And two years later, OPEC, Russia and Australia were listed by German analysts as having been the major obstacles on the road to the Kyoto Protocol.[38]

But while Australia continued to be seen as more of a petulant child than a team player in the aftermath of Kyoto, Hill was actually keen to see a market mechanism established, like an emissions trading system, that would be cost-effective in cutting pollution. He would not get very far. In the months following Kyoto, the sheen on his political star grew dull. In 1998 Howard established the Australian Greenhouse Office (AGO), a new executive agency of government, ostensibly to help the country meet its Kyoto obligations, and appointed a suite of ministers representing environment, industry and energy, foreign affairs, transport, agriculture and forests and fisheries to oversee its operations. According to Clive Hamilton, Howard's move was seen as a deliberate weakening of Hill's influence over emissions policy.[39]

Despite having been exposed to that prime ministerial kryptonite, Hill asked the AGO—as one of its first major undertakings—to prepare a series of discussion papers that would explore the design and application of an emissions trading scheme for Australia. The office responded quickly. A quartet of documents was released successively in 1999, covering design principles and an operational framework (March), permit allocation (June), carbon sinks (October) and monitoring and reporting (December).[40] However, little came from the effort. Emissions trading remained a concept without a champion in the prime minister's office.

Over the next several years, Howard waxed and waned on taking climate action and applying tools to address it. In 1999, for example, he introduced the $750-million "Measures for a Better Environment" package, the centrepiece of which was a program to pay big polluters to reduce emissions. In effect, as Guy Pearse describes, it was a program using taxpayers' dollars to buy "pollution cuts from big polluters."[41]

There were other greenhouse-gas policy measures during the Howard years. In 2004–05, his government announced it would fund steps to improve "emissions intensity," furthering, as Pearse puts it, "the pay-the-polluter" approach. Entitled "Securing Our Energy Future," the initiative also had some significant downsides, one of which was to wind down the Mandatory Renewable Energy Target (MRET) enacted just before the Kyoto Conference in 1997. That program had required electricity producers to generate an additional 2 per cent of their power from renewable sources by 2010. It was successful, and its demise was considered to be "a body blow" for the renewables industry.[42] Howard justified the move by suggesting that cheaper approaches were available elsewhere, pointing to emerging "clean-coal" technologies. However, clean coal was nothing but a chimera, a PR phrase dreamed up by the coal industry. An oxymoron then, it remains an oxymoron today.

CHAPTER 4

On June 5, 2002, Howard largely erased the memories of Robert Hill's triumphant return from Kyoto. It was World Environment Day, and the Kyoto Protocol had just been ratified by the EU and Japan. Kelvin Thomson, the Labor shadow minister for the environment, stood in Parliament and asked: "Given the Prime Minister's statement that the government is committed to meeting Australia's 2010 target for greenhouse gas emissions, isn't it in Australia's interests to ratify?"

Howard's reply gobsmacked much of the country: "It is not in Australia's interests to ratify the Kyoto Protocol," he said. Ratification "would cost us jobs and damage our industry."[43] The powerful fossil-fuel lobby had won the day, at least for the moment.

Howard's willingness to genuflect to big coal wasn't universally shared. Others in his government were not ready to throw in the towel. In the austral winter a year later, the Treasury, Industry and Environment departments, with support from at least six portfolios and the PM's senior office staff, proposed an emissions trading scheme. Guy Pearse said that this should have been "a sure bet," but that Howard "quashed the proposal" after consulting with industry.[44] The lobbies for coal, gas, aluminum smelting and mining industries—the "Greenhouse Mafia" as Pearse called them—won again, reinforcing a growing public sense that Howard was sacrificing the future of the country in both environmental and economic terms.

Gwen Andrews understood what had been happening in the backrooms. Appointed the first chief executive of the Australian Greenhouse Office in 1998, she resigned in 2002, later telling the *Age* newspaper that not once in her four years with the AGO had she been asked to brief the prime minister on the climate change issue.[45] *Not once.* All on its own, her reflection fully captures Howard's abdication of national leadership on the climate file.

Andrews went on to discuss the rot that had taken root in the Howard administration. "In my view, large energy, mining and resource interests in Australia had a disproportionate effect on government policy-making with regard to energy and climate change." She said the arguments yet again pitted environmental interests against economic interests, which were influential on government ministers. Bob Hawke's environment minister Ros Kelly had rebutted the environment-or-economy argument more than a decade earlier, but such specious logic still found a secure home in Howard's administration in 2002.

Good ideas are hard to keep down, however. Putting a price on carbon emissions through emissions trading—a central expectation of the Kyoto Protocol—is one of them. The Australian public understood that. A Newspoll survey in 2001 "found that 80 per cent of Australian adults supported ratification of the Kyoto Protocol," and only 10 per cent were opposed.[46] An ACNielsen poll in 2006 explored further, finding that 63 per cent of Australians were prepared to pay more to cut emissions, through both higher taxes and higher prices for goods.[47] Indeed, public support for reductions only grew in recent decades: as of 2016, 90 per cent believed that the federal government should take action.[48] That Howard could be so impervious to public opinion in the early 2000s and fail to understand that the issue would only strengthen with time reinforces the power of the "Greenhouse Mafia" lobby during his time in office.

In his final three years as prime minister, Howard pushed the nuclear option. Although "Securing Our Energy Future" in 2004 had ruled out the use of domestic nuclear power, Australia has very large uranium reserves. Howard was keen to see them exploited. In a perverse twist of logic, some of his front-bench MPs argued that if Australia exported a lot of uranium, it could claim it as a carbon credit—and in essence claim to be carbon-neutral.[49] In a snippet of cross-equator irony, it's not a big jump from that position Down Under to claims by BC Premier Christy Clark years later that by exporting vast volumes of liquefied natural gas—the production of which generates a large volume of carbon pollution—the province would be helping to combat global warming.

Despite strenuous efforts, Howard couldn't convince his own cabinet that nuclear had a big future. Senior ministers felt it was "politically dangerous" and urged avoidance.[50] Howard did find an ally in US President George W. Bush, who was quite willing to discuss how the two countries could collaborate on a nuclear-energy future and reduce emissions. That cross-Pacific interest in uranium did little, however, to counter domestic pressure to put a price on carbon. By 2007, as Guy Pearse put it, "no one in the business community believed emissions could be free forever."[51]

The Australian states announced their own proposal for a nationwide carbon-trading scheme to start in 2010.[52] And even though some of his own ministers had become gymnasts, backflipping on the nuclear issue and now supporting him, Howard had painted himself into a corner.

He could not keep rejecting the inevitable: carbon pricing was coming and big players in the world economy were already adopting it. The European Union had begun trading carbon permits within its member countries in 2005, establishing an enormous market and setting a price on carbon dioxide emissions. In California, the world's sixth-largest economy, Republican Governor Arnold Schwarzenegger signed a bill in September 2006 requiring market mechanisms to reduce greenhouse gas emissions.

In December 2006, Howard responded, creating the Prime Ministerial Group on Emissions Trading under the chairmanship of Peter Shergold, the top civil servant in the country.[53] Of the twelve members of the group, seven were senior executives from the business sector, and five—including Shergold—were secretaries of government departments.[54] All twelve were in favour of carbon pricing. The

Shergold report was released at the end of May 2007 and Howard wasted no time in accepting its recommendations. Behind Labor in the polls, and with the prospect of a looming federal election, he announced on June 4 that his government would introduce an emissions trading system (ETS) for Australia by 2012. It was a massive shift.

John Howard took his new position into the federal election campaign of November 2007. But was he intellectually comfortable with it? Apparently not, as he revealed six years later. On November 5, 2013, he delivered a lecture to a well-known climate-skeptic group in London, the Global Warming Policy Foundation. "I have always been something of an agnostic on global warming," he confessed, before going on to explain why he went into the 2007 election with emissions trading as a key policy initiative: he cited a "perfect storm" of an extended drought in eastern Australia, an early bush-fire season, the Stern report, and Al Gore's film *An Inconvenient Truth*: "To put it bluntly, 'doing something' about global warming gathered strong political momentum in Australia." Even years after the fact, Howard still appeared to be unable to connect the dots between the incontrovertible science of global warming and the marked climatic changes that were afflicting his country. He had considered "doing something" only because it was politically expedient, not because it was the right thing to do.

The 2007 election campaign was John Howard's last run for power. His years of dithering, of ducking the challenge, of offering rhetoric without action, of transforming Australia from international leader into international pariah were soon to end—at least temporarily. A new and extraordinary political roller-coaster ride was just about to begin, one that would stand in sharp relief against the climate-action blueprints quietly being drafted in British Columbia that very month.

CHAPTER 5

Down Under II: Yo-yo Politics and Australian Carbon Pricing
The Rudd-Gillard Years

> *We should be at a stage now in this country where climate change is beyond politics.*
> KEVIN RUDD, in a speech at the National Climate Summit, Parliament House, Canberra, August 6, 2007

BRITISH COLUMBIA'S POLITICS HAVE A RICH HISTORY. THOSE IN THE province might even go a little further and describe them as colourful or even outrageous. Fans of history remember that William Smith, BC's second premier, changed his name to Amor de Cosmos, albeit well before taking leadership of the province in late 1872. Carrying an appellation that translates as "lover of the universe" presented no impediment to holding high office in BC back then. Should you mention Amor de Cosmos east of the Rockies today, you'll typically be met with an eye-roll or a dismissive shrug—body language that implies many in Canada still view politics and politicians in British Columbia as being in some sort of left-coast Twilight Zone. Legislating a carbon tax in 2008 was seen by many easterners as just another example of BC wackiness.

But here's a bit of a secret. Wacky politics in the world of climate policy are not unique to BC. Some of the political nonsense that arose in British Columbia, like the NDP's "axe the tax" campaign, pales in comparison with what went on in Australia in the decade following the John Howard era. BC premiers were never toppled because of their positions on climate action, only to have them bide a bit of time, pop back up and reclaim the premiership. No leading politician in BC accepted climate science earlier in his career only to ignore it later, when it became a political inconvenience. A senior BC cabinet minister never held up a lump of coal in the legislature during a week of record-high temperatures and roared at the Opposition, "This is coal, don't be afraid, don't be scared"—as happened in the Australian Parliament on February 9, 2017.[1] The years following the Campbell climate legislation didn't have the sorts of political machinations that have wrapped Australia's carbon pricing history in a blanket of yo-yo politics: on/off, up/down, proclaim/revoke, legislate/repeal.

When BC's carbon tax experience is compared with what happened in Australia over nearly the same time frame, in 2008–13, the contrast could not be more antipodean. Where BC's carbon tax established a template for the world, Australia's multiple attempts to put a price on emissions in the post-Howard years yield a tale replete with lies, deception, backroom manoeuvring, political backstabbing, media misinformation, and the destruction of the political careers of at least one good man and one good woman.

Kevin Rudd won a decisive victory over John Howard in the election of November 24, 2007, taking eighty-three seats for Labor in

CHAPTER 5

the 150-seat Parliament. Rudd had become leader of the Opposition just the previous December, after serving as shadow foreign minister for foreign affairs. He was articulate, quick on his feet, relatively youthful at just fifty years of age, and had the advantage of being the new kid on the block. And in 2007 he was seen as a principled politician. Howard had been in office for eleven years—four terms—and was leading an increasingly fractious and tension-ridden Liberal/National coalition.[2] He was then sixty-seven years old, and many Australians saw him as yesterday's man. Rudd seized on the differences between the two, framing the campaign as "a choice between the future and the past."[3]

The two met in debate only once, on October 21 in the first week of the short thirty-nine-day campaign. It was a clear victory for Rudd. On the climate file, Rudd immediately lashed out, saying Howard "had no plan for the future." And when Howard promised that, if re-elected, his government would establish a carbon-trading program in 2011, Rudd pointed to the previous eleven years of inaction, saying:[4] "I don't believe, at the end of the day, he is really committed to this." Two years, later, Howard himself would admit as much.

Rudd had another advantage; he understood well what was happening on Australia's landscape. He'd been raised on a farm in Queensland about a hundred kilometres north of Brisbane. He says, "If you've grown up on a farm, you develop a deep appreciation for changes in the climate."[5] Asked what later drove him to take action, he points to the influence of his children and says that he began to read climate science himself. "The science spoke with such clarity," he says. "Those of us in the political process had a responsibility to our countries, and the planet, to act."

At the time that Howard-Rudd debate was held, the nation was eleven years into the fifteen-year-long Millennium Drought that afflicted the country from 1996 to 2010. Perhaps the worst drought of the last thousand years, the densely populated, vote-rich southern cities of Perth, Adelaide, Melbourne, Hobart, Canberra, Sydney and Brisbane were all hit hard, as rivers and reservoirs literally dried up. In 2003 the Australian Broadcasting Corporation's prime-time science program *Catalyst* aired a segment that interviewed some of Australia's top climate scientists.[6] They confirmed what the public could see all around them: winter rains were progressively disappearing. An already dry continent was becoming even drier.

The program went on to explain the science: the combination of regional ozone loss over Antarctica, which was cooling the upper atmosphere because less solar ultraviolet radiation was being absorbed, was combining with global-scale greenhouse warming to pull closer to the South Pole the high-altitude westerly winds that spin clockwise around Antarctica. The result of this pairing, both caused by human activities, was a shift of cold fronts to the south. Those fronts deliver winter rains, and as the narrator that evening pointed out, they "don't need to shift very far to completely miss the Australian continent."

All national polls in 2007 indicated that the public increasingly understood the link between global warming and the Millennium Drought, and wanted the government to take action. As leader of the Opposition, Rudd had put a firm hand on the issue six months before the debate. Along with the Labor-controlled state and territories governments, he'd commissioned Ross Garnaut, professor of economics at Canberra's Australian National University, to "conduct an independent study of the impacts of climate change on the Australian economy."[7] It was part of doing his homework. When he spoke of the need to reduce emissions and put a price on carbon to get there, his words had gravitas. In contrast, Howard's history of dithering on climate change finally caught up to him—when the issue arose, he was the debater with no clothes.

Howard had another albatross around his neck that October: his coalition was divided on his unwillingness to sign the Kyoto Protocol, and the divisions were very public. Michelle Grattan in the *Sydney Morning Herald* reported a week after the debate that "the Howard government has been deeply embarrassed by the revelation that Environment Minister Malcolm Turnbull proposed that Australia should ratify the Kyoto Protocol in a bid to defuse the issue electorally."[8] Turnbull was minister of the environment that austral spring. According to Grattan, behind closed doors some weeks earlier, he'd made the proposal to Howard's cabinet, only to have it rejected when the cabinet decided "a turnaround would not be credible after Mr. Howard had argued so strongly against ratification for so many years." Where Turnbull saw a glint of electoral gold that spring in attempting to mute his party's Kyoto history, his colleagues saw only dross. Turnbull would go on to play a pivotal but ultimately unconsummated role in Kevin Rudd's attempts to establish carbon pricing in Australia. In 2007, however, he accepted that addressing climate change was a must. Malcolm Turnbull was, in journalist Philip Chubb's words, "a tormented but genuine advocate of action."[9]

In what was described as "the world's first climate-change election,"[10] Rudd's Labor Party took 53 per cent of the two-party preferred vote[11] on November 24, 2007, enough to deliver eighty-three of the 150 seats in the House of Representatives. It did not do as well in the seventy-six-member Senate, where forty seats were up for grabs.[12] Labor took eighteen, a four-seat increase, to give it a total of thirty-two, five fewer than those held by the Liberal-National coalition. The Green Party took five seats, and others the remaining two. Labor's failure to gain an absolute majority in the Senate—such majorities being rare for either major party in the Upper House of Parliament—meant that Rudd was going to have to find seven Opposition senators to support his policies, if he was to get legislation passed. That turned out to be both pivotal and crippling, via a pathway that no one could foresee in the heady days after the election victory.

Rudd had promised during the campaign that as prime minister he would ratify Kyoto, a position he took in "the face of Howard's obdurate refusal to do so."[13] Moreover, his timing was perfect: record-high temperatures and the impact of the Millennium Drought were being reported in the media almost every day. And when a broad national poll held eight months before the campaign reported that 76 per cent of Australians thought climate change was a "major problem" and 77 per cent said they'd be willing to pay more to reduce emissions, Rudd knew that he had everything to gain and very little to lose. He wasted no time. Moments after being sworn in on December 3, 2007, and with the approval of his executive council—Deputy PM Julia Gillard and Governor-General Michael Jeffery—he added his signature to the instrument of ratification. "I did it simply because I believed it to be right," he said.

In the eyes of the global community, that signalled Australia was back, almost ten years to the day after Robert Hill had negotiated his last-minute Australia Clause in Kyoto. Rudd also announced that Australia "would establish a national emissions trading scheme by 2010."[14] He could not have guessed that this would indirectly sabotage his prime ministership fewer than three years later.

Two months later, Ross Garnaut released his interim review on the impacts of climate change in Australia.[15] He'd been given broad terms of reference, which included reviewing options for mitigating emissions. Garnaut's key recommendation was clear: Australia needs to establish a market for emissions trading and address "weaknesses and failures in related markets."

Kevin Rudd saw that recommendation as a road map to responsible climate policy. Just six months later, in July 2008, his government released a mammoth 530-page consultation document, a "Green Paper," that described Rudd's proposal, the Carbon Pollution Reduction Scheme, or CPRS.[16] The Green Paper proposed a cap-and-trade scheme that would provide "a strong incentive for business to cut carbon pollution." The keyword in the CPRS proposal was *business*. There was, at that point, no intention to target fossil fuel use by individual Australians, a point that stands in sharp contrast to BC's carbon tax, introduced that very month.

The proposed final design for the CPRS, in two remarkably detailed volumes, was released the following December.[17] Cap-and-trade "will put a price on carbon in a systematic way throughout the economy," promised the document. And by setting a limit on emissions by Australian businesses and then gradually ratcheting it down, "the right to emit greenhouse gases becomes scarce—and scarcity entails a price."

In theory, cap-and-trade programs have an advantage over carbon taxes, in that they set the level of pollution that can be emitted by a given jurisdiction. Carbon taxes cannot offer such certainty.

But there is a flip side. Overseeing a cap-and-trade system is bureaucratically onerous. Emissions must be reported and audited, an undertaking that enriches carbon-accounting firms while driving up costs. Competitive auctions must be held on a regular basis, with appropriate fiduciary control. Almost unavoidably, businesses will lobby to be granted exceptions or free permits and try to profit from the system, as Gordon Campbell warned, and as happened in Europe following the establishment of the European Emissions Trading System (ETS). The globally respected Brussels-based NGO Carbon Market Watch reported in 2016 that the over-allocation of free emission allowances to energy-intensive corporations mainly in Germany, France, Spain, Italy and the UK—permits that were subsequently sold at ETS auctions—was a major factor in the companies' windfall profits of over €24 billion ($35 billion Canadian) between 2008 and 2014.[18]

Ross Garnaut was agnostic on tax versus cap-and-trade in 2008, noting that in his report he was "very even-handed about the two." He remains so today, despite allocation and trading missteps seen elsewhere, largely in Europe. The government came down on the emissions trading side in the CPRS because that's where the whole discussion in Australia had been focused, he says.[19] "That's what was in the public mind, in the business mind, in the bureaucracy's mind. It would have been a major disruption ... to change horses."

Kevin Rudd introduced the Carbon Pollution Reduction Scheme bill on May 14, 2009,[20] twenty months after he'd stood at the podium at the National Climate Summit and offered what were to become oft-cited words: "Climate change is the great moral challenge of our generation. Climate change is not just an environmental challenge. Climate change is an economic challenge, a social challenge, and actually represents a deep challenge on the overall question of national security."[21] Nearly two years later, he was walking that talk.

But the bill he introduced differed significantly from the provisions laid out six months earlier. He'd decided to delay start of the CPRS by twelve months, to July 1, 2011, a step taken to manage the impacts of the 2008–09 global recession. Permit prices were to be fixed at $10 per tonne for one year, after which the trading market would kick in. And in a critical adjustment, the emissions reduction target specified in the White Paper (5 per cent by 2020 relative to 2005 levels) was ramped up to 25 per cent by 2020, a concession to the Australian environmental movement, which had been demanding greater ambition. Rudd built in wiggle room, however: that commitment would be met *only* if the world agreed to stabilize levels of carbon dioxide equivalent in the atmosphere to 450 parts per million by 2050. That caveat would come to haunt him.

The CPRS was much more complex than British Columbia's carbon tax. Designed to cover about 75 per cent of the country's emissions, the scheme was to apply only to big emitters, those spewing more than 25,000 tonnes per year of carbon dioxide into the atmosphere. That group comprised just one thousand of the 7.6 million registered businesses in the country. There was to be no direct carbon price applied to individual consumers, but it would be expected that the costs of the CPRS, such as increases in the price of electricity—largely generated by burning coal—would be passed on to ratepayers. Moreover, liquid fuel importers would be required to purchase permits that would drive up the cost at the pump of gasoline and diesel. Rudd planned to counter that fiscal imposition through commensurate reductions in fuel-excise taxes for the first three years.

Some praise and much criticism poured in. Don Henry, then CEO of the Australian Conservation Foundation, described the proposed legislation in positive terms as "a soft start," a position supported by the Australian Industry Group.[22] Others, quite predictably, were unhappy. On behalf of its member coal- and metal-mining industries, the Minerals Council bleated that "thousands of jobs" were at risk. The council was, of course, wrong. The CPRS would have stimulated employment in the emerging renewables industry, as in BC when the cleantech sector surged after the carbon tax was introduced. Job-switching in Australia, rather than job loss, could have become the name of the game, but the Minerals Council couldn't see past the end of a shovel. Moreover, the plan was to award millions of free permits over several years to the large coal-fired generators and other energy-intensive industries. Those entities would be able to sell excess permits if they improved their efficiency—for example, by switching progressively away from low-carbon-value brown coal or investing in new gas-fired generation. There was money to be made and new, cleaner jobs to be created, but parochialism yet again stood in the way.

Any optimism generated by the introduction of the CPRS in May 2009 was dashed on the rocks three months later. After 144 amendments had been debated and included, the House of Representatives passed the package on July 4 and sent it off to the Senate. On August 13 the wheels fell off the bus—for the first time. Coalition and crossbench senators defeated Labor's amended CPRS by a 42 to 30 margin. All five Green Party senators voted *against* the scheme, disgruntled that it failed to commit to an *unconditional* reduction in emissions of 25 per cent below 1990 levels by 2020. "They wanted to look purer than Labor in the eyes of environmentally aware voters," wrote Julia Gillard, then deputy PM, in 2014.[23] For Kevin Rudd, it was back to the drawing board.

Malcolm Turnbull, leader of the coalition Opposition in the House of Representatives, played a formative role over the next four months. He'd been in favour of an emissions trading system at least since serving as environment minister in the John Howard administration in 2007. Will Steffen, professor of earth system science at the Australian National University, served as science advisor to the ministry at that time, and describes Turnbull as someone who both "understood the climate issue very well and was very much for price-based action on climate change."[24] Rudd was amenable to negotiating with Turnbull—he needed the support of the Liberal senators if the CPRS was to pass. At Turnbull's insistence, the CPRS was revised to include large increases in compensation—essentially subsidies—for energy-intensive industries, as well as several billion dollars of support for the manufacturing and electricity-generation sectors. On October 22, the CPRS was brought before the House for the second time. On November 16, it passed third reading and was sent, again, on to the Senate, backed on this occasion by Opposition Leader Malcolm Turnbull.

Fearful of passage through the Senate failing a second time, given Labor's minority seat-holding and despite Turnbull's position,

CHAPTER 5

Rudd and Turnbull turned to Senator Penny Wong, then minister for climate change in Rudd's cabinet, and Queensland MP and shadow resources minister Ian Macfarlane, Turnbull's climate-change spokesman in the House. Macfarlane was known to be skeptical about man-made climate change. But he was also a Turnbull loyalist and commanded respect within the mining industry. He'd slowly come to accept that humans were playing a greater role in changing Earth's climate. Wong and Macfarlane were asked to hammer out an agreement on the CPRS that would later be approved by a Senate majority. They agreed on November 23 that the government would table amendments to the CPRS that would primarily increase compensation to industry by some $6 billion.[25] Bipartisanship was in the air, albeit at a cost.

Behind the scenes, the atmosphere was distinctly unsettled. Seeing Turnbull as a quisling on the climate file, the right wing of his coalition—a motley collection of deniers, fundamentalists and some conspiracy theorists—had been not-so-quietly laying the groundwork to oust him as leader. Their positions were laid out clearly on national television, in a September episode of the ABC *Four Corners* series entitled "Malcolm and the Malcontents." Nick Minchin, a lawyer, Liberal senator from South Australia and Opposition leader in the Senate, was a lead agitator. The group's knives were out. Minchin said onscreen that climate change provides the left with "the opportunity to do what they've always wanted to do, to sort of de-industrialize the Western world," adding that "they embraced environmentalism as their new religion."

Rupert Murdoch's newspaper the *Australian*—normally unsupportive of climate science in its editorial pages—reported that Minchin "came across as a 'complete fruit loop' when he suggested climate change was a left-wing conspiracy."[26] But Minchin's comments were perfectly in character; over a decade earlier, he'd been known for questioning "the addictiveness of nicotine and the harmfulness of passive smoking."[27] He was far from alone. His National Party colleague, Queensland Senator Barnaby Joyce, said in the *Four Corners* program that his constituents see climate change as "socialist chardonnay rubbish that is dreamed up by people who obviously have no real risk of having to pay for it," a position with which he clearly agreed.

Lurking behind both of them was Tony Abbott, the MP for Warringah in New South Wales. Abbott had infamously described climate change science as "absolute crap" at a public meeting on September 30.[28] But he was cagey—at that point, he wasn't entirely opposed to acting on it. Two months earlier, he'd been interviewed on Sky News.[29] He didn't appear comfortable when responding to a question about climate change. But after offering two denier talking points—that "climate change science is far from settled" and that "we've had cooling temperatures over the last decade notwithstanding carbon dioxide emissions"—Abbott went on to support, in general terms, what Gordon Campbell had done in British Columbia a year earlier. "If you want to put a price on carbon, why not just do it with a simple carbon tax?" he asked. "Why not just ask motorists to pay more; why not ask electricity consumers to pay more, and then at the end of the year you can take your invoices to the tax office and get a rebate for the carbon tax you paid? It would be burdensome ... but it would certainly raise the price of carbon without increasing in any way the overall tax burden." Probably unknowingly, he was roughly channelling BC's simple and effective revenue-neutrality approach. He went on to say, "This emissions trading scheme is almost impossible to understand ... I think there'll be all sorts of avenues for scamming, but this is what the government wants. We can't counter from Opposition." At that time, he had no clearly defined position. He was unwilling, at least on-air, to put a stake in the ground.

CHAPTER 5

That changed four months later.

Like Abbott, Nick Minchin was a shadow minister under Turnbull. On November 25, the sun arced across a clear blue sky in Canberra. By 4 p.m. the temperature had reached 34° C on the broad lawns outside Parliament House, some fourteen degrees above normal. Minchin, Abbott and three other front-benchers—all of whom were global-warming skeptics—chose that abnormally hot afternoon to resign their positions, ostensibly in protest at Turnbull's position on the CPRS.

Two days earlier, Turnbull had taken the Wong-Macfarlane deal to his Liberal MPs for approval, but at a wild eight-hour-long meeting he ran into a wall of backbench opposition. Philip Coorey, writing in the *Sydney Morning Herald*, said that Turnbull was "rolled" at the meeting. He'd made one serious mistake, presuming that he could count on his twenty-member shadow cabinet to support the deal, enough to give him a three-vote majority in the party room, or so he thought. But that was not to be; of the seventy-three MPs at the meeting, forty were opposed to the deal. Turnbull refused to yield and called off a formal vote for approval, arguing that, regardless, the CPRS had to go through the Senate as planned that month. "We're going to vote for it," he said, an insistence that infuriated many coalition MPs who were opposed. In the end, the combination of denialism and Turnbull's intransigent position precipitated the resignations of the "Minchin Five." Turnbull's leadership was now directly in peril. Tony Abbott was asked if he planned to challenge for the leadership of the Liberal Party. He declined, but added: "I can't say what might happen in the future."[30]

It didn't take long for the future to arrive. Two days later, Abbott—whose natural political fit was with the party's hard right wing—announced that he would challenge Malcolm Turnbull for the Liberal leadership. Front-bencher Greg Combet, then minister of defence and science in the Rudd government, describes in Technicolor the Opposition's thinking: "The real conservatives, like the Trump-type people in the Liberal Party, realized, 'We don't care if this climate change stuff is real or bullshit. What's important is that it's a political opportunity for us to take control of the Liberal Party because we can convince our base ... [that] we'll be able to get rid of some fandoodling city hepcat like Malcolm Turnbull."[31]

On December 1, that "fandoodling hepcat" faced Abbott's challenge in the party room. Standing before the Liberal MPs, Turnbull famously said, "I will not lead a political party that's not as committed to effective action on climate change as I am." Moments later, he lost to Abbott by a single vote. Greg Combet says that vote "completely transformed the previous decade of politics on this issue." One of Australia's most respected senior reporters, David Marr, described Abbott's ascension in words spare but grave: "The bipartisan alliance on global warming was over."[32]

The yo-yo years in Australian politics had begun.

The timing of the mass resignation had not been coincidental. The Senate had commenced debating the CPRS for the second time on November 17. By November 30 it was anticipated that the final vote in the Senate chamber would be called on December 2, the last sitting day before the Christmas break. Abbott knew that if Turnbull lost a leadership challenge before then, Liberal senators would be less likely to vote in favour of the revised CPRS, despite the Wong-Macfarlane agreement. He was right.

Positions held by the various parties became immediately clear on November 17. Recognizing that the Wong-Macfarlane negotiations were ongoing, Senator Bob Brown, leader of the Green Party, groused that senators were being asked to "embark on a debate about legislation which is far from finalized."[33] His unhappiness with the deal-making approach was on full display. The Senate was "being treated as a necessary routine to go through, but not the equal second house of the Parliament, which is the watchdog of the people's interests," he complained. Many commentators pointed out that his umbrage was rooted not so much in the details of process, but rather in being personally snubbed by Rudd. "The Greens were ignored," noted former reporter Cathy Alexander.[34] She figured Rudd was avoiding getting close to the Greens because he thought they were a threat and would "cannibalize [votes] from the left."

The debate continued for several more days, alternating between demands for unconditionally greater national ambition and recitation of tired denier memes from the usual suspects: "trees will not grow without carbon dioxide," "the climate has been changing for a hundred million years," "scientists trained in this area cannot make up their minds," "Australia's emissions are insignificant at the global scale," "the Great Barrier Reef will adapt," "thousands of jobs will be lost." The Greens' Bob Brown moved an ambitious amendment on November 18.[35] He insisted again, as he had the previous August, that the bill needed to commit *unconditionally* "to reduce emissions by at least 25 per cent below 1990 levels by 2020" and express "a willingness to reduce emissions by 40 per cent below 1990 levels by 2020 in the context of a global treaty." While his exhortation was entirely justifiable, based on the best science, it was not well-received by the Labor members more closely attuned to political feasibility.

By November 30, the debate had almost run its course. Tensions were high. Senators were irritable and spoke in increasingly fractious terms. Senator Chris Evans, Labor minister for immigration and citizenship, rose in the chamber and accused Minchin and the deniers of being willing to "tear the whole [Liberal] party apart rather than accept action on climate change."[36] He pointedly added that Minchin's "treachery has been on display all week" and accused him of "being duplicitous with your own leader."

Bob Brown had a different complaint, grumbling that the Opposition arranged with the government "to transfer an extra $7 billion or $8 billion to the polluters," while reminding the chamber that Kevin Rudd had failed to reply to his letters that fall, which laid out "what this nation should be doing."[37] He accused the prime minister of being "disdainful ... of Senate process."

Senator Penny Wong delivered searing criticism of the Liberal Party, accusing it of abandoning the national interest.[38] "The Senate has been a sideshow for the internal divisions in the Liberal Party," she said, "whilst they try to get the numbers in their own party room, including attempting to tear their leader down."

That evening, Senator Christine Milne, deputy leader of the Green Party, returned to the major objection raised earlier in the day by Bob Brown. Referring to the enhanced support for industry agreed to in the Wong-Macfarlane pact, she said that the CPRS presented a "completely unjustified wealth transfer—ripping it out of the pockets of Australians and into those of the coal-fired generators. There has never been a more disgraceful example of unjustified policy."[39] And on the penultimate day of debate, December 1, Milne accused the government of running up "the white flag of political expediency," saying that it did "not have the political courage to actually tell people the truth about the fact that we are in an emergency."[40] Milne implied that the CPRS would lock in failure. With that sentiment hanging in the thick air of the chamber, Senator Wong shot back: "I say to the Green Party: locking in failure is voting with the extremist right, the extreme climate-change deniers in the coalition ... against action on climate change."

The next day, December 2, the Canberra sky was more grey than blue. Bob Brown captured the historic significance of the occasion as he delivered his final speech on the CPRS. "At stake here is the future of this nation, which is more vulnerable to climate change than any other nation on earth," he said.[41] But even with that future pressing so firmly on his conscience, he was unable to get past his concern that the CPRS was more of a sellout than a solution. "I know the Prime Minister has said that we must act for our children and our grandchildren," he said, "but the question is: do we act according to the world's brain trust or do we act according to the pressure of the big polluters?" Casting his eyes over the Opposition benches, he lamented that the Great Barrier Reef—so seriously threatened by global warming and ocean acidification—was "being put to sacrifice by this mob of sceptics."[42]

Yet Brown was still poised to vote no. The CPRS wasn't enough for the Greens. His colleague, Christine Milne, reinforced that position, saying: "We have a proposition from the government that the Carbon Pollution Reduction Scheme is action on climate change. It is a fake claim. It is a fraud for anyone who understands the climate science."[43]

Penny Wong closed the debate. "I will just say this. This legislation may well fail on the Greens' vote and, whatever rhetoric those Green senators engage in, they will have voted for Australia's carbon pollution to continue to rise. They will have voted against action on climate change ... We should leave this place being able to look Australians in the eye and say: 'We acted. We took responsibility.' Instead, some of those opposite will simply have to look Australians in the eye and say: 'I voted this way. I voted for the future not now and in fact not ever.'"

The vote was called at 11:52 a.m. Two Liberals crossed the floor to vote for the CPRS. But in a dramatic statement, all five Green senators voted against, defeating the legislation by an eight-vote margin. Had the Greens supported the CPRS, Australia would have had an emissions trading system: imperfect, yes, but a start. Rudd would have been able to report to the international community that Australia had stepped up to the plate. But none of that was now to happen.

Looking back and lamenting that historic failure, one senior civil servant said simply that "the policy wasn't pure enough for the Greens."[44]

In their zeal, the Greens sought purity and ignored practicality, failing to heed the aphorism that "the perfect is the enemy of the good." But Kevin Rudd must also share the blame, for it was he who was averse to inviting the Greens into the fold and working to accommodate their concerns, as pure as they might be.

Twelve days later, Rudd flew off to Copenhagen to lead a one-hundred-member Australian delegation at the UN-sponsored conference on climate change. According to the *New York Times*, he'd "hoped to present Australia's emissions trading system as an example of the developed world's willingness to commit to binding targets."[45] But the failure back home had taken the bloom off that rose—it had harmed his credibility on the global stage. Despite his best efforts to act as a broker between developing countries and the developed world, the Copenhagen meeting failed to live up to expectations.

Rudd arrived back in Australia "broken in spirit," according to Julia Gillard.[46] He was also in a bit of a political pickle. During his absence but with his agreement, acting PM Gillard had announced that the government would reintroduce the CPRS legislation for the *third* time in February 2010. In her words, this "would give the Liberals one more chance to support it." Rudd thus had two choices: he could stick with that plan, or he could invoke a double dissolution, or "double D," a constitutional procedure unique to Australia wherein the prime minister requests the governor general to dissolve both houses of Parliament and call a full election. A double D can only be triggered if (or when) the Senate and House of Representatives fail twice to agree on a piece of legislation, as happened on December 2. And there was a third, unspoken option: the government could walk away, at least temporarily, from taking serious climate action. It was not to be an easy choice.

Rudd dithered on going for the double D. "It's always a hard call," Greg Combet says, noting that the failure to reach an agreement in Copenhagen had made it even harder. But Rudd had the constitutional justification to proceed, and the overthrow of Malcolm Turnbull meant the Liberals were now in disarray. Combet is convinced that Rudd would have won an election if it had been held in early 2010. "The issue was in really sharp contrast. Rudd still had a good degree of community support," he says. "Hindsight tells you we should have done it, and we would have had a legislated emissions trading system that would have been legitimate."

Kevin Rudd's hindsight is different. "Only one of twenty members in the cabinet supported going to a double dissolution on the CPRS," he remembers. He was not willing to roll the dice. Former journalist Cathy Alexander thinks there was another issue. "If [Rudd] had gone to a double D, he might have ended up with a whole swag of Greens in the Senate ... and that would have totally locked him into having to get every single piece of legislation through [them] in the Senate. That's one thing he was worried about."

But more broadly, the Australian public was confused. It couldn't understand why moving to a double D wasn't a high priority if climate change was indeed the "greatest moral challenge of our generation," as Rudd had famously said over two years earlier. "There was a jarring in the public," said one senior civil servant. Don Henry agrees. With his background as former head of the Australian Conservation Foundation and later a public policy fellow at the University of Melbourne, Henry had an innate sense of the pulse of the nation. Looking back, Henry says that Rudd would have had a very good chance with the double D because of the drought: "The issue was very tangible in people's everyday lives."[47] Henry suggests that Rudd had just the two options then: "Either take it to an election, or keep prosecuting it and keep working hard to change the Senate into a majority position."

Rudd and his team temporarily settled on the latter. On February 2, 2010, the CPRS—now formally incorporating the amendments agreed to in the Wong-Macfarlane pact three months earlier—was again introduced to the House. It was the first sitting day of Parliament that year. Many of the speeches in the following week rehashed contentious ground and some quite predictable full-on nonsense. Stupidity was on full display. In his lengthy contribution, Western Australian Liberal MP Dennis Jensen asked: "Carbon in its natural state is graphite. Is that what the government is attempting to ban?"[48]

Against that backdrop of dross, one Liberal MP's speech stood out. Twenty minutes after Dennis Jensen had finished offering foolishness, Malcolm Turnbull rose from his seat to deliver an impassioned and wise twenty-seven–minute oration.[49] He began with a reminder about responsibility: "All of us here are accountable not just to our constituents but also to the generations that will come after them, and after us." He recognized that the CPRS would benefit business, saying: "An Australian emissions trading scheme, with a carbon price set by the market, would improve business investment certainty." Moreover, he noted that the amended CPRS "appropriately balances environmental effectiveness and economic responsibility." And in a point that would have resonated with Gordon Campbell and his finance minister, Carole Taylor, Turnbull reminded his fellow MPs that the CPRS would be fair, noting that the scheme "is not designed—nor should it be—to raise additional net revenue for the government, as taxes do, since the funds raised by the sale of permits will be returned to compensate lower income households and assist businesses, especially those which are emissions-intensive and trade-exposed and cannot readily pass on the increase in energy costs." He ended by saying "I will be voting in favour of this legislation."

Three days later, the CPRS passed third reading in the House, and eleven days after that it was, again, put forward for first reading in the Senate. If Yogi Berra, the king of déjà vu, was looking down on Canberra, he must have been smiling.

It took only moments for fireworks to erupt in the Senate on February 22. The coalition Opposition, now led by Tony Abbott, proposed a stalling tactic, asking for the package of CPRS bills to be considered one by one. Victoria Senator Kim Carr—Labor minister for innovation, industry, science and research, who had introduced the bills moments earlier—exploded.[50] His words of frustration richocheted off the Senate walls, calling the proposal a ruse and "clearly a measure of the hypocrisy of the leader of the Opposition. He says he wants to be the straight shooter. He says he wants to be the plain man's politician. What we have here are old-fashioned Liberal Party tricksters." He went on to warn the Liberals that "we will be calling you the hypocrites and frauds that you are."

It was not a good start.

On February 24, second reading was moved in the Senate. But the government knew it did not have the votes to push ahead. The Greens, still seeking purity, were intransigent in their opposition. Malcolm Turnbull's willingness to cross the floor in the House and support the bill had little influence on Liberal senators who now saw more political value in aligning themselves with Tony Abbott. Worse, Rudd himself appeared to have lost interest. Journalist Philip Chubb wrote in 2014 that Rudd's post-Copenhagen focus had shifted to hospital reform, while on the climate file he was increasingly paralyzed. Chubb claims that Rudd's demeanour at the time was seen to be "agitated and angry,"[51] and cites Wayne Swan, the Labor treasury secretary, as saying, "Whenever he didn't know what he wanted to do, he just didn't do anything."[52]

Meanwhile, smelling political blood and with the Murdoch media empire as an ally, Tony Abbott was taking pains to repeat endlessly a devastatingly simple description of the CPRS.[53] It was "Kevin

Rudd's great big tax on everything," he said. Rudd seemed unable or unwilling to counter. Toward the end of February, the CPRS morphed into the political equivalent of petrified wood. It was still there in the Senate, texture and grain intact, but it had lost any pretense of having life. By failing to go the double-D route and call an election, Rudd had ceded the one big lever he had: the prospect of "an early election to resolve a political impasse."[54] With the failure of Copenhagen in the rear mirror, and Abbott slowly shifting public enthusiasm away from carbon pricing, Rudd had essentially given up. It was a time in which, according to Don Henry, "a lot of policies became roadkill to pretty ugly politics."

The most crippling bombshell was yet to come. In March, Kevin Rudd was actively considering ways to announce surrender on the CPRS. He commissioned an internal briefing paper that proposed delaying the scheme to 2013, subject to international consensus being reached. But he never went public with this plan, "because he feared the scorn of voters," according to Philip Chubb.[55] By mid-April, budget decisions had to be finalized. Rudd was under increasing pressure to make a final decision. On April 21, the cabinet agreed to take the CPRS out of the budget.[56]

Six days later, on April 27, the rug was pulled out from under Rudd's shuffling feet—someone leaked to the *Sydney Morning Herald* that the CPRS was dead. But Rudd had not yet publicly admitted the surrender. In taking questions from journalists later than morning, cornered after giving a doorstop speech at a hospital, he was skewered by his own words.[57] "A little while back you said that climate change is the greatest moral, economic and social challenge of our time. With this now being delayed, do you still believe that to be the case?" one reporter asked, before following up with a tripartite knockout punch: "What about Australia being a world leader though? So we're now waiting for the rest of the world? What happened to the idea of us leading the way?" Nothing Rudd might have said could have relieved the devastation of that moment as the television cameras rolled.

That scrum shattered what was left of Rudd's credibility. He had fought the 2007 election on the overwhelming need for climate action, and was now seen to have cut and run. The "leader had fled the field," wrote Philip Chubb.[58] For the Australian Labor Party, the CPRS had become the symbol of political catastrophe. Someone had to pay. The yo-yo was about to spin again.

By late June, amid plunging poll results, and with Tony Abbott scoring points in the media, the stage was set for drama within Labor. On June 23, Deputy PM Julia Gillard walked down the hall and asked Kevin Rudd for a ballot. It was a declaration that she was challenging him for the leadership of the party. The election was held the next day: Gillard won unopposed after Rudd withdrew his candidacy. Australia had its first female prime minister, a historic moment, albeit one achieved more by default than by design.

Could Rudd have done things differently? Could he have emulated Gordon Campbell's success with carbon pricing in British Columbia? After all, Campbell had introduced his tax a year before Rudd first introduced the CPRS to Australia's Parliament—*and* Campbell had been re-elected in BC ten months after the tax took effect. Campbell had chosen the bureaucratically simple route of a broad-spectrum, revenue-neutral tax, while Rudd, politically fearful of the T-word, had gone with the less wieldy CPRS. But both had shared one key design feature: carbon-pricing revenue was to be recycled to the less well-off in their respective jurisdictions.

Why did Rudd fail so dismally where Campbell succeeded? There are five key reasons.

First, Campbell was a conservative with an overwhelming majority in the provincial single-house legislature—he had no skittish

Senate to appease. Rudd was the opposite: a democratic socialist with all that implies, and although Labor had a majority in the House, it had to rely on other parties in the Senate, some of whom were openly hostile to carbon pricing in any form.

Second, Campbell was a decisive leader. He consulted, listened, deliberated and acted. He didn't duck hard decisions. Rudd was a study in contrast, plagued by procrastination when the going got rough.

Third, British Columbia is rich in hydropower; some 92 per cent of its electricity is supplied by water rushing through turbines. Australia is starkly different, with four-fifths of its electricity generated by burning coal. The act of pricing emissions from coal-powered plants was unavoidably going to drive up the cost of electricity for all consumers, including industry. The political implications of that were, and are, obvious.

Fourth, British Columbia had no equivalent of a Tony Abbott nipping at the heels of government like an obnoxious yapping dog, calling global-warming science "absolute crap" and misguidedly describing carbon pricing as a "great big tax" without mentioning that revenue would return to those who needed it most.

And fifth, although in BC media ownership is concentrated in a few hands, the province had no direct equivalent of Rupert Murdoch's dominant editorial control opposing climate science and carbon pricing.

But of all these, arguably the most important was leadership. Rudd was unwilling to prosecute his case by calling for a double D, even when his closest advisors said the public still supported serious action on climate. Paul Kelly, editor-at-large of the *Australian* newspaper, wrote that Rudd, "despite having branded climate change the greatest moral test of the age," was plagued by "a failure of conviction."[59]

CHAPTER 5

Gordon Campbell made a different choice. He won the provincial election in 2009 "because of our position on climate," he says.

But like the good idea that just won't go away, the carbon pricing opportunity was not to disappear in Australia. It was soon to come back, this time wearing a different set of clothes.

CHAPTER 6

Down Under III: Yo-yo Politics and Australian Carbon Pricing
The Demise

We've seen everything go wrong with climate policy in Australia in the last ten years. I never thought we'd end up in such difficulties with respect to climate policy in 2017.
    CATHY ALEXANDER, former journalist with Associated
        Press Australia

You'd have to have a few kangaroos loose in the top paddock to consider life as an Australian politician.
    SIMON KENT, *Toronto Sun*, June 26, 2013

WHEN JULIA GILLARD ASCENDED TO PRIME MINISTER OF Australia in late June 2010, the news hurtled around the world. In Toronto, it was given a less than congratulatory Canadian spin by the editors of *The Globe and Mail*, who wrote: "In banana republics, coups d'état often occur when the leader is out of the country. In Australia, the conspirators opted not to wait a day. Instead, Kevin Rudd was dispatched by his deputy the day before he was to depart for the G20 meeting in Toronto."[1] Across the pond—well, the Tasman Sea—the New Zealand media took a different perspective. The NZ Herald chose to focus more on the forthcoming Gillard-Abbott duel, using six stark letters in giant bold type, above the fold, to make their point: "GAME ON."[2] No exclamation mark. No emphasis necessary. They were Gillard's two words, whispered into Tony Abbott's ear as he shook her hand just after she was sworn in.

Despite her whispered challenge, it was no game. In the wake of Kevin Rudd's unwillingness to go to the wall on carbon pricing, she knew she had to reclaim Labor's credibility on the file. At her first press conference as PM, she announced that if her party won the forthcoming election, she would "re-prosecute the case for a carbon price at home and abroad."[3] Politically savvy, she added a caveat that gave her wiggle room: "I will do that as global economic conditions improve and as our economy continues to strengthen."

On that very day, British Columbia's simple, revenue-neutral carbon tax had been in place for almost exactly two full years. It was already working. Since the tax began, per capita fuel consumption in the province had fallen by 8 per cent relative to the rest of Canada. But Gillard remained unwilling to consider a similar approach for Australia—she would advocate only emissions trading. Tony Abbott's "great big tax on everything" campaign had worked: he'd poisoned the well. *Carbon tax* in Australia was a toxic phrase, even while it was meeting with considerable success in British Columbia.

Gillard wasted no time in seeking a mandate. After just twenty-four days as PM, she called on the governor general and requested an election be held on August 21. Game on: Julia versus Tony.

It was a furious five-week campaign. All seats in the House of Representatives and half the Senate were up for grabs. Abbott hit hard and early, accusing Gillard of being untrustworthy, having offed Kevin Rudd as leader. At stop after stop on the hustings, he used *trust* as a verbal shiv. The irony must have escaped him, given his actions in spilling Malcolm Turnbull months earlier. But five days before the election, Gillard didn't help perceptions of her own trustworthiness. Asked by a television reporter about her position on a carbon tax, she stared straight into the camera and said, "There will be no carbon tax under the government I lead."[4] It was a truthful statement. But she had stopped short, failing to add something she had been consistent about during the campaign: she *was* prepared to legislate a carbon *price*, just not a *tax*. The absence of *price* and *emissions trading*—as she puts it, the other half of the answer—was to come back and bite her less than a year later.[5]

In his excellent book *Power Failure*, Philip Chubb offers a detailed critical analysis of Gillard's electioneering during the campaign. It "was a disaster," he wrote.[6] Knives were still out within the Labor ranks. There were leaks from Kevin Rudd's camp, reporting on confidential cabinet discussions in which Gillard was reported to take oppositional positions on issues such as pension reform. Chubb says these revelations undermined Gillard's authenticity with voters.[7] In the end, Labor lost eleven seats in the House. Abbott's Liberal-National Coalition picked up seven. They were tied, seventy-two seats apiece in the 150-seat chamber, the first hung Parliament since the early years of the Second World War.

It was no better for Gillard in the Senate. Labor lost three more seats, giving it thirty-one, eight short of a majority. The big story was the Greens' success; they picked up one seat in the House of Representatives and an additional four senators, giving them a nine-member Senate caucus. For the first time, Australia's Green Party held the balance of power in both parliamentary houses.

Gillard and Abbott were both determined to form government, but Labor held the better ideological hand. The Greens would have been hard-pressed to form an alliance with a Liberal-National leader who had earlier described climate science as "absolute crap." The Greens faced another constraint: they had to wear their repeated history of demanding purity. Gillard recognized that the Greens were no longer in a position to insist on an economically unpalatable emissions-reduction target (even when a severe target is clearly necessary if global warming is to be limited to 2° C). She perfectly captured that recognition in her memoir, writing: "Pride in purity easily becomes the ignominy of impotence."[8] In the end, Gillard won the support of three independents and the sole Green MP in the House, and all nine Greens in the Senate. On September 14, she and her cabinet were sworn in, holding the reins of power by just a single seat.

The Greens' support came at a cost. Greg Combet—then newly appointed as minister of climate change and energy efficiency—says, "We were a minority government, and the Greens said, 'If you want to get stuff through, there's a price ... We want you to take action on climate change and we want it to be serious.'" It was a price Gillard was willing to pay. Her formal confidence-and-supply agreement with the Greens read, in part: "Reducing carbon pollution by 2020 will require a price on carbon." Privately, she hoped to establish an emissions trading scheme by mid-2012, "provided agreement could be struck quickly enough."

She didn't delay, immediately creating a multi-party committee that included herself as chair, Greg Combet, Green Party Leader Bob Brown and four others. Climate scientist Will Steffen and economist Ross Garnaut were appointed to provide expert advice. "We explored every bloody option, from a carbon tax to trading schemes to subsidies," Combet says, "and arrived back—not surprisingly—at an internationally linked emissions trading scheme." But it was not smooth sailing. "I very much wanted a floating price scheme [with Europe], because that was a defensible argument with business to say 'Now [with] our second-largest market, we've got the same carbon price.'" Agreements were negotiated, and there was a treaty with the EU, "but the Greens said, 'No, we want a fixed price,' effectively a carbon tax for a period of time. Gillard and I had to make a fixed-price period at the front end of it: a carbon tax for three years and then a floating price scheme. It was most unfortunate, but that's all I could negotiate ... It still gives me the shits, to be honest, because it was a destructive thing that the Greens made us do."

Politics is the art of compromise. By February 2011, five months later, Gillard's minority government had a draft design for the Emissions Trading System (ETS). It was poised to work. Gillard had herself conducted the final negotiations with Bob Brown, who, as Green leader, had been so vehemently opposed to Kevin Rudd's CPRS two years earlier because it didn't go far enough. Brown was then sixty-six years old. Stung by the failures of the past, he had become more politically pragmatic—Gillard described him as being in "legacy mode."[9] In the end, the two leaders settled on an initial price of A$23 per tonne, a figure judged to be similar to the likely 2015 price in the European Union's ETS.

Mere days later, a bomb landed. Rupert Murdoch's flagship newspaper, the *Australian*, ran a story that leaked the basic structure of the deal. It prematurely forced the government's hand. On February 24, Julia Gillard and Bob Brown met in the parliamentary courtyard to explain the pact to the media. But by then the anti-tax genie was out of the bottle and in full flight. Thanks to the three-year fixed-price agreement, the right wing was gleefully able to conflate emissions trading with a carbon tax, presenting them as direct equivalents.

In Question Period that afternoon, Tony Abbott was first to rise. His parliamentary seat was upholstered in rich green leather, but the symbolism of that colour was far from his mind. "My question is to the prime minister," he began.[10] "I refer the prime minister to her repeated promise before the last election, including on the very day before the election, 'I rule out a carbon tax.' How can she justify today's betrayal—" He stopped abruptly, drowned out as members on both sides shouted slurs. As the Speaker demanded order and the hubbub slowly receded, Abbott continued, only to be cut off as his time expired: "How can she possibly justify today's betrayal? If the Australian people could not trust her on this, how can they trust her on anything?" Julia Gillard shot back, both barrels blazing, accusing Abbott of seeking to "destroy and wreck and spin and mislead." It was open warfare.

That evening, Gillard made it worse. On ABC Television's *7.30* program, commentator Heather Ewart pushed Gillard on the word *tax*. She asked nearly the same question three times, and on the third go Gillard appeared to have relented.[11] Ewart asked, "With this carbon tax—you do concede it's a carbon tax, do you not?" Gillard replied, "Oh, look, I'm happy to use the word *tax*, Heather." With that resigned reply, she'd stepped into the proverbial cow-pie, mightily. *Tax*, that politically toxic word, was now in the middle of the table.

Philip Chubb describes that Thursday as "the day from hell" in *Power Failure*, spawning "a public campaign of intimidation by business, media, and coalition opponents."[12] In his words, it "escalated into a fiery crusade that was devastating for the prime minister's community standing and for the government's prospects for re-election in 2013."[13] Gillard's own words were being used to pin her with one of the most powerful words in the lexicon of politics: *liar*.

But fiery crusade and hair-splitting semantics aside, Gillard had the votes, thanks to the Greens. The emissions trading scheme passed both houses of Parliament on November 8, 2011, embedded in the comprehensive Clean Energy Bill 2011.[14] After several years of starts, stops and false hopes, Australia finally had legislated carbon pricing. It was to take effect the following July 1, on the very day that British Columbia raised the quantum of its carbon tax to its then apex price of C$30 per tonne, with hardly anyone in the province batting an eyelash.

Gillard's Clean Energy Bill had two key features beyond emissions trading, both designed to recycle ETS revenue across the economy. The first addressed industry's concerns. Some $2.5 billion in free permits was to be distributed to energy-intensive, trade-exposed Australian businesses to help them with the transition. The coal-fired electricity sector was central; Gillard says Greg Combet "worked painstakingly through the support necessary to the electricity sector to ensure continuity of supply."[15]

Equally important, she and Combet recognized that the ETS afforded an opportunity to reform Australia's income tax structure. It provided assistance to those who would be hit hardest by forthcoming increases in energy prices, particularly for electricity. That reform was extraordinary and far-reaching, a near-parallel variation on what Gordon Campbell had insisted on in BC: revenue received by the Australian Treasury from the sale of emissions permits was to *triple* the tax-free threshold, taking it from A$6,000 to A$18,200 in 2012–13 and A$19,400 in 2015. In a statement that was music to the ears of lower-income earners, Greg Combet's ministry reported that "people with incomes below the new taxfree thresholds will get to keep all of their wages in their regular pay packets."[16] Some one million low-paid Australians were removed from the tax rolls, thanks to the boost in the threshold. Moreover, the changes in the tax code—along with provision of higher Family Tax Benefits, pensions and other support programs—were designed collectively to balance the increased cost of energy for households. It was smart policy, and it was there at Julia Gillard's insistence.

Six years later, over coffee at a Melbourne hotel, Greg Combet—now back in the private sector—sat reflecting on the campaign to establish the ETS. He was still steamed about Tony Abbott's mendacity in 2011, telling me: "He would have called black 'white,' you know. That's the sort of politician he is." In his memoir *The Fights of My Life*, Combet took a harder line, writing: "Abbott's behaviour was beyond the pale. I wondered how he could live with such calculated political dishonesty."[17] At every opportunity in Parliament, says Combet, Abbott used the carbon price to launch attacks on the prime minister that were "always intensely personal, belligerent and ruthlessly reductionist."[18] Through it all, Rupert Murdoch's right-wing press "amplified his claims as part of a campaign for a change of government."

But a belligerent Abbott, even with the "cynical political animus" of the Murdoch media empire on his side, could not halt progress.[19] Gillard's legislation worked. World Bank data show that from 2010, the year before the Clean Energy Bill took effect, to 2013, per capita carbon emissions in Australia fell by over 3 per cent.[20] Greg Combet was emphatic on that success, noting that emissions in the electricity sector went down, and that brown coal, the most emissions-intensive, lost market share. But "no one went out of business. It was clearly driving change. You can see it in the emissions results."

In his book, he lists other positives: over 155,000 new jobs were created in the first year of the ETS and the economy grew by 2.5 per cent (which, in another parallel, is almost identical to the 2.4 per cent rate of growth enjoyed by British Columbia over the same period, when the carbon tax there was in full flight at c$30 per tonne).[21] And yes, electricity prices did rise, by about 10 per cent, a jump fully expected, but that increase was due in part to transmission improvements, not just the emissions tariff. The higher costs were offset for consumers and industry, thanks largely to tax-code changes and transitional government funding. Moreover, the sale of emissions permits raised A$3.8 billion in its first six months, money that flowed back into the economy to support tax cuts, energy efficiency and renewable energy initiatives, and assistance for carbon-intensive, trade-exposed industries. There was a lot to like.

As clear as the success was, however, it didn't register politically. Abbott continued to rail against the "great big tax on everything." "He outmanoeuvered us on the politics profoundly," Combet says. Kevin Rudd didn't help. One senior government insider says the former PM was playing a destructive game inside the Labor government, slyly criticizing Julia Gillard while discreetly angling to take over the leadership once again. It was a dangerous combination, akin to Abbott's subterfuge of Turnbull two years earlier.

The polls showed Labor losing ground. In a stroke of pure irony, a change in climate reinforced that decline. The Millennium Drought broke. As the billabongs refilled, the stark daily reminder of a drier future faded quickly from consciousness, at the same time as the electricity charges and natural gas prices were rising. Liberal (i.e., conservative) state governments zeroed in on the price jumps, requiring "electricity retailers to put a huge red stamp on your electricity bill," says Combet. "It was like someone put a stamp on there saying, 'You know the Gillard government's carbon tax has cost you $45.62 this billing period.'" And just like what had happened in British Columbia, the compensating cuts to income tax were forgotten. Australians saw a big red stamp on their electricity bill but were blind to the bigger take-home on their pay stubs.

By the austral autumn of 2013, Julia Gillard's polling statistics were dismal. To be fair, other issues were eating at the population. Prominent among them was the sporadic but heavily publicized arrival of boatloads of illegal immigrants from South Asia. Australia struggled to cope with that influx while disenchantment with Labor grew. Moreover, the Minerals Council of Australia ran a highly successful anti-carbon-tax campaign that effectively influenced public opinion. In concert with the Coal Association, the council had "used the media to advance their interests, in the knowledge that some agenda-setting outlets would treat them uncritically," according to Philip Chubb.[22]

Bare-faced misogyny played a role too. In the 2012 Human Rights and Social Justice Lecture at the University of Newcastle, bestselling Australian author Anne Summers—who once ran the federal Office of the Status of Women when Bob Hawke was prime minister—said that Gillard had been "attacked, vilified or demeaned in ways" specifically related to her gender, noting that "Tony Abbott is a serial offender."[23] But she also implicated Kevin Rudd, suggesting he had "successfully struck at her credibility and her authority" with "cruelly targeted leaks." Talk radio in Australia was also particularly vicious. Equating it with open-line programs in the US, Greg Combet says there was a lot of "vomit coming out: 'She's a bitch. She's a liar. She's a backstabber.'"

On June 24, 2013, Newspoll—which described itself as "Australia's leading public opinion polling company"—released its latest assessment. It was crushing. "Labor is headed for a landslide defeat and Tony Abbott has Gillard's number," wrote Dennis Shanahan, political editor of the *Australian*.[24] Abbott "has turned a nine-percentage point deficit against Kevin Rudd as preferred prime minister three years ago into a record twelve-point lead over Gillard," the paper reported, suggesting that, among other issues, Gillard's public persona had been damaged by "the broken promise not to introduce a carbon tax." Abbott's dissembling—without a doubt aided by the Murdoch press—had won the day: two of three Australians said they would choose him over Gillard, according to Newspoll.

Two days later, yo-yo politics struck again. Gillard—hearing rumours that her own MPs were questioning her leadership, and fed up with the internal squabbling—called a leadership vote. It was the third such internal Labor Party vote in 2013. The previous two, one in February and one in March, had been requested by Kevin Rudd, and Gillard had won both, handily.

This one was to be different.

"I have been in a contest with the leader of the Opposition, but I've also been in a political contest with people from my own political party," she told BBC News on June 26.[25] She was referring directly to Rudd, who quickly announced he would stand for the leadership, for the third time in five months. He was flip-flopping on a promise he'd made just after losing a challenge in March, after which he'd issued a statement that said: "Mr. Rudd wishes to make 100 per cent clear to all members of the parliamentary Labor Party ... that there are no circumstances under which he will return to the Labor Party leadership in the future."[26]

Gillard laid down her own challenge that afternoon: "Anybody who enters the ballot tonight should do it on the following conditions: that if you win, you're Labor leader, that if you lose, you retire from politics." It was game-on once more, but this time with a different opponent and a bigger price to be paid. A few hours later she lost, fifty-seven to forty-five.

The political career of Australia's first female prime minister was over.

Commentators everywhere took notice, many marvelling at the blood-sport nature of Australian politics. Canada's then doyen of political analysis, Jeffrey Simpson, suggested in *The Globe and Mail* that Gillard's loss represented a bit of a reckoning: "It was a swift, brutal, ruthless change—just as it had been three years before when Ms. Gillard stuck a knife in Mr. Rudd's political back and mounted a successful coup against him. He was out, she was in, and he never, ever forgot."²⁷

Kevin Rudd was now at a crossroads. The Australian political Mixmaster had spun him back into the prime minister's chair, just in time to launch an election campaign. He knew if that if he was to hang on as PM for three more years, he'd have to act on the climate issues that continued to dominate news cycles. He also knew he had to divorce himself from Gillard's carbon policy. And so, on July 16, just three weeks on from reassuming the chair, he called a press conference at which he announced that the government "has decided to terminate the carbon tax" by moving up the date of commencement of the open-market emissions trading scheme to July 1, 2014.²⁸

Julia Gillard's pact with the Greens, which guaranteed the three-year fixed price, was dead. Rudd rationalized the shift by citing cost-of-living pressures and impacts on small business. The cost of carbon emissions then fell by almost $20 a tonne to match the EU trading system price. A whiff of hypocrisy wafted through the air as Rudd claimed, "Most Australians want to see their nation doing its bit, playing its part to protect the environment from the effects of climate change." Cutting the price on emissions was no way to protect the environment, but it was a way to curry political favour before an election.

Rudd's announcement came with a big federal budget hit: some A$3.8 billion in forgone revenue over the next four years. The treasurer in Rudd's new cabinet, Chris Bowen, spun that as best he could, claiming that hundreds of millions would be saved by bringing forward free emissions permits to 2014–15 "and discontinuing the program after that." Fewer dollars would be available for the Carbon Capture and Storage flagship program, the biodiversity fund and the Carbon Farming Futures program. And to top it all off, the Australian public service would reduce the number of senior executives by 1 per cent.

It was not a good day for senior civil servants or climate action in Australia.

What's worse is that the Gillard plan had been working.[29] In the twelve months since it had begun, emissions from brown-coal combustion fell 13 per cent, electricity production from renewables was up nearly 30 per cent and the economy had grown by 2.5 per cent. Had he been asked, Gordon Campbell would have expressed no surprise at these statistics, for BC had witnessed similar responses after the province's carbon tax took effect: the economy strengthened and emissions fell. But Rudd's rejection of carbon pricing reflected influences that British Columbians were fortunate to never experience: political mendacity, right-wing media malfeasance and a climate denial campaign funded by Big Coal.

In the end, Kevin Rudd's Hail Mary was all for nought. On September 7, 2013, Tony Abbott swept to an overwhelming victory, his Liberal-National Coalition capturing over 53 per cent of the two-party preferred vote and taking 90 of 145 seats. While personally devastating for Rudd, the political swing was a shattering setback for climate action. Abbott wasted no time in addressing the carbon pricing issue. "In three years' time, the carbon tax will be gone," he proclaimed to raucous cheers, as television cameras panned across his supporters' faces during his acceptance speech. Many of those faces, maybe most, were young, their euphoria misguided. They either didn't know or didn't care that, unlike the few old-timers in the room—John Howard among them—they had a lot to lose. The future took a blow that night.

Abbott was sworn in as prime minister on September 18. The very next day, now the leader of a country in the midst of its hottest year and hottest September in recorded history, Abbott introduced legislation to repeal Julia Gillard's Clean Energy Act. The public had been primed. Abbott's coalition had conducted alchemy: the phrase

CHAPTER 6

clean energy had been magically transmuted to *carbon tax*, two words synonymous with poison in the minds of many.

That magic wasn't accidental. Aussie pollster Alex Frankel pulled back the curtain of semantic manipulation in 2015, writing about how Liberal strategy architects had shaped public sentiment away from the real issue—global warming—and toward the toxic T-word:[30]

> *Every time the words 'carbon tax' came out of Abbott's mouth, he was delivering on this strategy to steer debate away from the inconvenient truth about Australia's huge coal industry ... What was most striking about this strategy was how the Liberals managed to execute it without any effective opposition. Most of the Parliament, the media and society bought into the language of 'carbon' and 'carbon tax,' letting the Liberals frame the debate.*

The strategic philosophy underlying the campaign was simple: control the narrative and you control policy. On that same first day of governance, Abbott's administration moved not just to repeal carbon pricing, but also to put bridle and reins on the narrative: new environment minister Greg Hunt rang Tim Flannery at 10 a.m. that morning to tell him the climate commission had been abolished.

Flannery, the bestselling author and scientist, had been appointed head of the commission by the Gillard government in 2011. It comprised a group of leading scientists and business gurus, tasked with providing independent, non-prescriptive public information on the science and effects of global warming. Flannery saw its primary role as being "purely educational." In just over two-and-a-half years, it developed into an intensely admired organization seen to be apolitical, both within and outside Australia. Flannery was heartened by the commission's progress, saying: "It left me with a really profound faith in the common sense of ordinary people when they are given the facts. We had quite a number of people who came into our events either skeptical or hostile, who ended up in [a] different composition when they left, which was good, because they could see that we weren't just green ratbags."[31] But Abbott and his right-wing minions saw the commission's work differently. It was a threat to the anti-carbon pricing narrative. It had to go.

Neither Abbott nor Hunt could have anticipated what was to happen mere days later. As a public call for the immediate reinstatement of the climate commission swept across Australia, the newly sacked Tim Flannery sat down with Amanda McKenzie, a senior communications advisor. McKenzie had co-founded the Australian Youth Climate Coalition in 2007 and served as its national director for four years before joining the commission in 2011. She was savvy; she knew how to engage youth, more than a hundred thousand of whom had signed up to join her coalition. Flannery wanted to know about crowdfunding: Did McKenzie think it was feasible that the now defunct climate commission could rise by attracting public donations? The idea was to create a new but essentially identical publicly supported body, one with a similar name: the Climate Council. "I knew nothing about crowdfunding or any of that," Flannery says. "Amanda reassured me that ... what we were doing was ambitious [but] it was not impossible. It was within that framework that we said we would go ahead and do it."

The public response to the Climate Council's online launch at midnight on September 23—just four days after Greg Hunt had sacked Flannery—was gobsmacking. "We launched with a crowdfunder and within the first day we'd raised over $500,000. Within another [day] we'd raised over $1 million from roughly fifteen-thousand to twenty-thousand people all just giving $30 or $20," said McKenzie in 2015.[32] Flannery remembers those initial moments: "We got a $15 donation from a guy in Western Australia, our first. I think it was a minute after midnight. It was, like, 'Wow, that's great.' We had messages from people saying 'Look, I voted for Tony Abbott but I never expected this.'" The phoenix had risen, in part spurred by backlash. Tony Abbott had attempted to quash access to climate information of direct relevance to Australia. He failed.

The repeal of the carbon tax fared somewhat better, albeit not quickly. On November 13, 2013, Abbott stood on what was to be a raucous first day of Australia's forty-fourth Parliament. Sixteen heckling protesters were removed from the gallery as he personally presented the repeal legislation. "The election was a referendum on the carbon tax, the people have spoken, now it's up to this Parliament to show that it's listened," Abbott said.[33] He went on, offering standard neo-liberal anti-carbon-pricing orthodoxy, saying that the bill delivered "on the coalition's commitment to the Australian people to scrap this toxic tax. It is also a cornerstone of the government's plan for a stronger economy built on lower taxes, less regulation and stronger businesses."[34]

What Abbott didn't say was that repealing the tax but keeping in place the income tax reductions—and other fiscal supports that Julia Gillard had brought in two years earlier—would cost the treasury more than $3 billion a year. He ducked that reality in his speech by spinning it as a plus, his words lacking even a hint of the "responsible fiscal management" that conservatives so often trumpet: "The carbon tax will go, but the carbon tax compensation will stay so that every Australian should be better off."[35]

From a fiscal perspective, that was bad enough. But then it got worse. "The government will scrap the carbon tax and then proceed with its Direct Action Plan," Abbott announced.[36] The centrepiece would be "the Emissions Reduction Fund [ERF]—a market-based mechanism for reducing carbon dioxide emissions ... [that] means more trees, better soils and smarter technology—this is the right way to get emissions down," he claimed.

To be fair to Abbott, he wasn't the worst of the zealots in his ruling coalition, but he was, perhaps, the most skilled political opportunist. Throughout his parliamentary career he'd been careful to avoid *expressly* ruling out the need to reduce carbon emissions. Moreover, as leader of the Opposition, he'd backtracked on his widely cited public comment—made in 2009—that the science around climate change was "absolute crap."[37] During the 2013 campaign, while remaining overtly hostile to carbon pricing, he offered a surprisingly thick olive branch when asked about global warming: "I think that climate change is real; humanity makes a contribution. It's important to take strong and effective action against it, and that is what our Direct Action policy does."[38]

"Direct Action" had first been developed formally in 2010. The conservative coalition, then in Opposition, argued that relative to the year 2000 this policy would lower emissions 5 per cent by 2020 without taxing compensating households or burdening industry. In his parliamentary speech, Abbott was using the phrase *market-based* loosely, for the ERF was in essence a rather weak voluntary program through which applicants would compete through reverse auctions for grants—read: subsidies—supplied by the government. The cheapest plans were thus most likely to be funded. Grants would, for example, support the planting of many millions of trees, pay farmers to avoid clearing vegetation from scrubland, offer rebates for the installation of residential solar photovoltaic cells, promote carbon storage in soils, and pay industry to reduce its emissions.

It was a perversion of Adam Smith's famous metaphor: under Direct Action, the invisible hand of the market was to metamorphose into highly visible handouts.

The program attracted many critics, some internal. Malcolm Turnbull—then a Liberal backbencher—had previously denounced it in scathing terms: "Industry and businesses, attended by an army of lobbyists, are particularly persuasive and all too effective at getting their sticky fingers into the taxpayer's pocket," he told Parliament in February 2010.[39] It was a "recipe for fiscal recklessness," he said. Ben Eltham, a journalist at Australia's leading independent think tank, the Centre for Policy Development, pithily summarized Direct Action as a plan "based on incomplete science, dubious economics and breathtaking political expediency. It will be hugely expensive. It won't cut carbon emissions."[40] But nearly four years later, having brushed such critics aside, here was Malcolm Turnbull's nemesis, Tony Abbott, smiling while announcing to the country that Direct Action was back.

When the idea was first discussed by his own party in 2009, Turnbull described Direct Action as "a con, an environmental fig leaf to cover a determination to do nothing."[41] Frank Jotzo, a leading environmental economist at the Australian National University, more recently described it simply: it's "a subsidy policy," he says,[42] adding, in an echo of Turnbull, that the ERF was created "to be able to say that there is a climate change policy in place." Direct Action would keep happy "those pro-Liberal supporters that have at least a passing interest in environmental matters."

None of these complaints mattered to Abbott. Fiscal recklessness: *pfft*. Science that refuted the benefits of his soil-carbon sequestration program: *pfft*. Subsidizing dirty brown-coal-fired power plants: *pfft*. All that mattered to him was the pretense that his administration was acting on global warming. And for nothing more than maintaining that charade, he was willing to sacrifice Labor's Clean Energy Act.

Abbott's carbon-tax repeal passed in the House without amendment seven days after its introduction. But it was a pyrrhic victory, at least initially. Abbott had a problem: any bill passed in the House by his right-wing majority was vulnerable to being trumped in the Senate, where the Liberal-National Coalition held only thirty-three of seventy-six seats. The twenty-five Labor senators and ten Greens needed the support of just four of the remaining eight crossbench senators to vote down any legislation from the House. A handful of those eight were members of the Palmer United Party (PUP), a conservative splinter group founded in early 2013 by a billionaire mining magnate named Clive Palmer.

That splinter would turn out to be critical.

Clive Palmer is a rotund former real estate agent turned swashbuckling owner of golf clubs, coal mines and nickel refineries. He had been active in the political backrooms of conservative eastern Australian politics for four decades. He fit well the characteristics of a brusque, self-made, uber-wealthy right-winger, one for whom the word *diplomat* offered two syllables too many. After being suspended by the Queensland Liberal-National Party in 2012 for calling the deputy premier of Queensland a "thug," Palmer resigned and founded the PUP in 2013. It was marginally successful in the federal election later that year: Palmer eked out a bare majority in his House riding. More important, the PUP took Senate seats in Queensland and Tasmania and, in a special election in April 2014, a third seat in Western Australia.

A month earlier, Abbott's attempt to repeal the carbon tax had collapsed; the Senate narrowly rejected the legislation at its third reading on March 20. But that rejection was like lopping the head off a hydra only to see it regenerate. The bill, now with minor amendments, was reintroduced in mid-July. Green Leader Christine Milne condemned what was about to come.[43] A vote to abolish the carbon tax "is a vote for failure," she decried during the third-reading Senate vote. "This Parliament does not want to face up ... to what is intergenerational theft."

But the die was cast. The trio of PUP senators, reinforced by an informal alliance with a fourth—a member of the fringe Australian Motoring Enthusiast Party—stood behind Abbott's coalition and voted for repeal. The final tally in the Senate: thirty-nine ayes and thirty-three nays. Palmer's rump senatorial group carried Abbott's water that day. But the ten or so weeks leading up to the vote had not been without some rather peculiar shape-shifting drama.

On April 3, Palmer was interviewed by Tony Jones on the Australian Broadcasting Corporation's *Lateline*, its late-evening news program.[44] Jones asked Palmer point-blank: "Do you believe the consensus scientific view set out in the latest IPCC report that climate-change impacts due to global warming will have especially serious impacts on Australia?"

"No, I don't believe that's so. There's been global warming for a long time. I mean, all of Ireland was covered by ice at one time," responded Palmer, drawing on a well-worn denier theme. "I think that's part of the natural cycle." A minute later, he slyly implied that hundreds of the world's leading climate scientists might have been bought off, saying: "Well, I can get a group of scientists together, Tony, and pay them whatever I want to and come up with any solution. You know that's been happening all over the world in a whole range of things. There's a long history of that happening in the drug industry."

In making such comments, Palmer wasn't really rejecting science. Instead, his core concerns reflected his background as a coal mine owner. Referring to carbon emissions, he said, "Just look at industry, [and] not just take away our jobs, up our electricity prices. Let's care about the people living on the planet right now." Intergenerational theft was far from his mind that April evening, which happened to be exactly three months to the day since the Australian Bureau of Meteorology had reported that 2013 was the country's hottest year on record.[45]

In the following weeks, the shape-shifting occurred. Don Henry, the highly regarded former head of the Australian Conservation Foundation, brokered behind-the-scenes discussions between Palmer and Al Gore that culminated in a very public press conference in Canberra on June 25.[46] Palmer, the coal magnate, stood side by side with Gore—the world's most recognized climate activist—in one of the most unlikely pairings ever witnessed in the world of climate politics. Palmer announced that his party's senators would vote to abolish the carbon tax. That in itself was no surprise. It was his next pronouncements that stunned the assembled reporters.

"Climate change is a global problem and it must have a global solution," he proclaimed. "Australia needs to do its fair share." And then—as former vice-president Gore stood stoically—came the bombshells: "Palmer United senators will move an amendment to establish an emissions trading scheme ... and will not support any change to [existing] renewable energy targets."

In his complimentary following remarks, Al Gore avoided the cancellation of the carbon tax and, though likely deserving, he took no credit for the newly displayed metamorphosis of Palmer's position on climate action. Instead, he singled out the intent of Palmer's party "to re-implement an ETS" and "support the continuation of the Renewable Energy Target [RET]." In response to later criticism that the appearance had added legitimacy to a politician who didn't deserve it, Gore fought back, noting the RET was "the element of Australian policy responsible for most of the carbon reductions."[47]

Twenty-two days after what had probably been the most unusual press conference ever held in the world of Australian climate politics, a tiny party led by a coal-mining baron cast the deciding votes that rendered Australia the first country in the world to reverse action on climate change.

The carbon tax was axed.

It was instantly global news. Rupert Murdoch's *Telegraph* unsurprisingly gloated in the UK, running the headline "Australia abandons disastrous green tax on emissions," with the subtitle "Prime Minister Tony Abbott has finally won backing to end the tax on carbon emissions Down Under, signalling the pointlessness of such schemes."[48] Other outlets took a more objective view, with London's *Financial Times* noting that the repeal "threatens to isolate the country amid increasing international efforts to tackle climate change."[49]

Within Australia, a sense of embarrassment, even shame, was palpable. Michael Raupach, director of the Climate Change Institute at the Australian National University, offered a particularly accurate but sobering analysis: "The repeal of Australia's carbon price is a tragedy, not a triumph. It flies in the face of three giant realities: human-induced climate change, the proper role of government as a defender of the common good, and the emerging quiet energy-carbon revolution."[50]

The repeal marked the end of a long political road, one that ABC journalist Julia Baird captured perfectly, writing that it "defined three elections, destabilized three prime ministers and dominated public debate in this country for eight toxic years."[51]

How did Australia's carbon emissions fare in the aftermath? The short answer is: not well. The decline in emissions that gave Greg Combet such a sense of pride quickly reversed: a 10 per cent decline in carbon dioxide emissions from the electricity-generating sector between mid-2012 and mid-2014 flipped to a 3 per cent rise in the six months after the repeal.[52]

Australia's about-face stands as a travesty that Tony Abbott will forever wear. For five years he'd trolled through the muck that coats the denier community, repeating ad nauseum the phrases "great big new tax" or "axe the tax." But as it turns out, he wasn't alone in his stridency. Far across the Pacific, the same "axe the tax" cry was heard, surprisingly loudly, in British Columbia in 2009, just a year after Gordon Campbell brought his visionary policy into being. What was astonishing was that this time the anti-tax stridency came from the progressive side of the political spectrum.

EPILOGUE: *There is schadenfreude in Australian politics. In the months after the carbon tax repeal, and following a varied suite of policy missteps, Abbott's approval rating steadily declined, reaching a record low in February 2015.[53] On September 14 of that year, Malcolm Turnbull issued a challenge to Abbott's leadership. Late that evening, Turnbull won the spill motion, fifty-four to forty-four. He became, again, Australia's prime minister, the fifth to take the helm in six years. Abbott's term was the shortest such period of leadership since 1972, aside from Kevin Rudd's three-month-long incarnation in 2013.*

CHAPTER 7

Cutting Off One's Nose to Spite One's Electoral Prospects

The 2008–09 Campaign to "Axe the Tax" in British Columbia

There is, conventionally, no reason to be shocked or dismayed that your average politician might be tempted toward short-sightedness and expedience. But in Canada, the NDP *once had a reputation for putting principle ahead of political gain.*[1]

RICHARD LITTLEMORE, 2009

As THE PACIFIC OCEAN QUENCHED THE SETTING SUN JUST WEST OF British Columbia's capital city on May 29, 2008, NDP Leader Carole James stood in the legislature and voted nay on the third reading of Gordon Campbell's Carbon Tax Act.[2] Twenty-nine sitting NDP MLAs joined her. Some held such disdain for the tax that they "physically turned their backs on the vote," according to *The Globe and Mail*.[3] Repugnance aside, their actions were for nought, subsumed by forty-one government votes. The carbon tax, introduced and passed by a conservative government, was now to take effect at midnight, just thirty-two days later.

For the left-wing NDP, it was a portentous moment. Award-winning writer Richard Littlemore has accurately described the NDP as "a traditional coalition of social policy progressives, labour activists and environmentalists."[4] But that night, in their unanimous rejection of the carbon tax, the New Democrat MLAs were, perhaps unwittingly, snubbing their traditional coalition. They made a collective decision that ripped apart British Columbia's environmental-activist community, sundered the provincial NDP membership, caused concern within the caucus, and played a significant role in determining the outcome of the provincial election a year later.

Three months earlier, Carole James—by any yardstick a committed, thoughtful, and pragmatic progressive politician—offered a typically wise perspective about the need to put a price on carbon emissions: "I think a revenue-neutral carbon tax that really looks at supporting low- and middle-income families, that actually is phased in so people can manage, that provides them with options to make change—then I think it's worth looking at."[5] James, probably without foreknowledge, was describing with remarkable accuracy key elements of the BC carbon tax introduced just five days later.

But something changed in the months that followed. James's progressive appeal yielded to the siren call of populism.

By the middle of 2008, about a year before Tony Abbott began repeating the phrase in Australia, "axe the tax" appeared in the NDP's playbook. At a caucus retreat in mid-June, almost four months to the day after she supported carbon pricing on TV, James stood in the bustling Interior city of Kelowna to rail against the carbon tax. "In two weeks, Gordon Campbell is going to take record-high gas prices and make them even higher," she said.[6] Politically, her timing couldn't have been better. The day before, the price of gas in Vancouver had hit $1.48 per litre as it moved upward in lockstep with the world price for oil. A mere seven months earlier, Vancouver drivers had been filling the tanks of their Hondas, BMWs and Toyotas for just $0.84 per litre.

James was capitalizing on that stunning sixty-four-cent jump. The sharp rise had hit some corners particularly hard. Socially concerned progressive that she was, James zeroed in on one particularly important concern in her own riding: "In Victoria, Meals on Wheels has announced that it will be ending its service because of high gas costs."[7] But lost in the rhetorical shuffle was the fact that the carbon tax would add only 2.4 cents a litre onto the recent eighty-four-cent jump at the pumps. The 2.4 cents was just noise. Nevertheless, she said, "People are angry. They're angry at the government. They're angry at the premier in particular, for not recognizing the pressure that this puts on them." In conclusion: "It's time for Gordon Campbell to axe his gas tax."

Campbell was a convenient political target, but he was having none of it. Later that day, on the CBC, he described James's position as disingenuous. "There is a whole lot of confusion from the NDP. Friday, they're for a carbon tax. Now they're saying there shouldn't be a carbon tax."[8]

Eleven months before the next election, Carole James had thrown down the "axe the tax" gauntlet. It was a blatant attempt to find an issue that "could turn things around for the party," says David Schreck, well-known NDP strategist.[9] "What they need is something, a ballot box question, that can get voters steamed" by voting day.

That "something" was well-organized. On the very day that James made her call, a new website—www.AxetheGasTax.ca—went live. A petition entitled "Axe the Gas Tax" was posted on the NDP caucus site, right under the header "Carole James, Leader of the Official Opposition."[10] And Michael Smyth of the *Province* wrote: "You can bet the negative TV attack ads are already being prepared for the election."[11]

It was a strategy destined to become—in the eyes of many economists, environmentalists and a large swath of the progressive electorate—a spectacular fail. The revulsion was palpable. When asked how she felt when she heard of James's announcement, Simon Fraser University Professor Nancy Olewiler, one of Canada's leading environmental economists, replies without hesitation.[12] "I was livid. It was just shocking, in the sense of 'Are you crazy?' It was pure political pandering to the fear that [it was] a tax grab."

An academic expert in carbon pricing, Olewiler had been consulted by BC's Ministry of Finance as the revenue-neutral structure was being designed in late 2007. She knew every detail. The notion promoted by the NDP that the tax was a revenue grab was "despicable." She points out that the Campbell government cut personal and corporate income tax rates. "It was revenue-negative in the beginning, because those tax cuts came in on January 1 [2008] and the carbon tax came in on July 1—so they had six months of lower tax rates. I was apoplectic on the environmental side, but also on how they distorted the economics," she says, concluding: "I can't say how incensed we all were."

Professor David Green of the Vancouver School of Economics at UBC was similarly chagrined. Two years earlier, Green had been instrumental in publicly advocating for carbon pricing. Green lived in Gordon Campbell's riding on the well-heeled west side of Vancouver, and was a member of a local advocacy group, Voters Taking Action on Climate Change (VTACC). In mid-2007, VTACC participated in an informal constituency meeting, focused on how BC should or could deal with the climate challenge. Premier Campbell attended the discussion. "I said, 'Look, there's a much more effective way to do this, which is the carbon tax,'" says Green.[13] "I expected to have to explain it to him. Instead, he looked at me and said, 'Yeah, we're really interested in that.'" Green recalls that Campbell went on to say "We need somebody to be out in front of us on this. We're sort of interested in going this route. But politically, to just stick our necks out would be something hard. If we had somebody who is actually generating interest, that would be very useful to us."

Green and his colleagues saw that as a green light to build a groundswell of support for carbon pricing and give Campbell political cover without, he said, "ever actually endorsing the Liberals." In short order, he put together a letter calling on the BC government to enact a revenue-neutral carbon tax.[14] It was signed by "almost every academic economist" in the province, says Green. Not surprisingly, when Carole James took to the hustings with axe-the-tax, Green found it "so disgusting. It really bugged me."

That exasperation extended well beyond economist circles. Tzeporah Berman, one of British Columbia's most prominent and respected environmental advocates, remembers that the entire axe-the-tax campaign left her personally devastated.[15] "This was going to be [the NDP's] wedge issue with Gordon Campbell, and they ended up on the wrong side of it," she says. Nancy Olewiler concurs. "[For] a party that had stood on environmental issues, to go one-eighty [degrees] and just throw them all out the window to gain a few points? It backfired ... they absolutely were on the wrong page, the NDP."

University of Victoria climatologist and future BC Green Party Leader Andrew Weaver was publicly infuriated. Eight days after James's speech in Kelowna, Lindsay Kines of the *Victoria Times-Colonist* described Weaver as having "ripped the NDP for releasing 'inaccurate' information about the tax."[16] Weaver was quoted calling the NDP's behaviour "reprehensible," concluding: "I say shame on them."

What, then, was behind the NDP's stance? Was it just political gamesmanship, maybe even soul-selling, in an effort to win an additional seat or two? Clearly, Carole James was playing on public anger, accusing the government of piling a tax on top of what Kines described as "skyrocketing fuel prices." Although that may have resonated in some parts of the province, particularly in rural areas where travel distances and home-heating costs were issues, it didn't resonate with savvy progressives or green urban professionals. Andrea Reimer, future Vancouver deputy mayor and executive director of the Western Canada Wilderness Committee, saw through the political spin. "Unless there's something that the NDP hasn't told us yet about what's motivating their campaign, the only conclusion that we can draw is it's a very crass look at polling numbers and how they might benefit."[17]

Bill Tieleman agreed, at least regarding polling. Tieleman is a communications consultant in Vancouver with a long history of political activism. He's a consummate organizer blessed with an innate capacity to harness and massage public opinion and steer it toward end points that fit with what he—and perhaps he alone—would see as his progressive agenda. But Tieleman looks at the world through the lens of politics first—and then, and only then, through ideological spectacles. As a populist, he was particularly astute in recognizing the building resentment as the carbon tax neared its inauguration day.

On June 10, 2008, just a week before the NDP's anti-tax campaign was formally introduced, Tieleman launched a Facebook group, "Axe the BC Gas Tax." For him, it was a made-to-order issue. "You have to drive it and drive it mercilessly," he says.[18] "They have the opportunity to make Campbell wear not just the carbon tax, but the market increase as well." His campaign quickly took off, attracting thousands of members in short order.[19] Bill Tieleman was suddenly a folk hero, at least in the eyes of the anti-taxers venting on his Facebook page.

It was a particularly odd time in BC's odd political history, marked by strange bedfellows. Tieleman was a left-winger at heart, yet he found himself sharing a mattress with the Canadian Taxpayers Federation, a right-wing, anti-almost-all-taxes lobby group that, like him, saw the carbon tax as an unfair burden on society. He wears such associations like a badge of honour: "I had really hard right-wingers, very conservative folks, praising me," he says with a smile.[20] In his blog posts, Tieleman saluted northern municipalities, the Canadian Trucking Alliance, right-wing editorial boards and Vancouver Island NDP MLA John Horgan, all of whom climbed on the anti-tax bandwagon.

Years later, Tieleman remains proud of the role he played. When asked if he was the axe-the-tax architect, he says, "I guess, in the sense of a broader public campaign, yes, I started that," adding a caveat: "I was not the architect of the NDP campaign at all." He was "pleasantly surprised" when Carole James took his phrase and ran with it, but Tieleman says he had no contact with the campaign.

Tieleman's pleasure, however, was rooted in a misperception. He argues that all consumption taxes are regressive: "They inordinately punish lower-income taxpayers and have a much lesser effect on high-income taxpayers." But the architecture of BC's carbon tax—with significant personal and business income tax reductions *and* the ongoing provision of direct cash transfers to northern, rural and lower-income residents—was *not* regressive. Because of the fiscal transfers and income tax reductions—which for personal income taxes were applied at the bottom end of the upwardly accelerating tax structure—BC's carbon tax was progressive. Moreover, it turned out to be revenue-negative. More dollars were returned to BC citizens than were collected via the carbon tax. That's simply a fact.

Tieleman's narrow focus on the phrase *gas tax* was an emperor-without-clothes tell, illustrative of a populist Svengali's spin (to borrow Richard Littlemore's term)[21] to sell political wares. When it is pointed out to Tieleman, directly, that the "gas tax" is actually a broad-spectrum emissions tax applied to much more than gasoline, he replies, "Yeah, that's true." So, why seize on the phrase *gas tax*?

"Because that's what most people see it as. The overwhelming majority of people see it as a gas tax because they are paying it. When they fill up their cars, they see it."

The discussion continues: "But it's a little bit of a perversion of what the tax actually is. Would you agree with that?" "Well, I certainly talked about the other elements, the heating fuel tax," he replies. "But everybody saw it as a gas tax. To me, it was pretty straightforward it was a gas tax." No nuance was necessary in his mind: "Calling it a carbon emissions disincentive would not have been too popular—and I am a populist." He was confidently appealing to individual resentment: an astute, if slightly less than honest, political strategy. The fact that the cost to individual British Columbians was *at least* compensated by the cash-in-pocket design of the legislation was to him an incidental consideration, one he was, unfortunately, too happy to disregard.

Despite Tieleman's very prominent Facebooking and Carole James's campaigning, something happened over the course of '08 that pushed the "axe the tax" campaign into the background. The world oil price collapsed, from a high of nearly US$170 in June to about US$51 per barrel at the end of the year. The price at the pumps plummeted correspondingly. The phrases *gas tax* and *axe the tax* largely disappeared from BC's political vocabulary.

That all changed in April 2009. A writ was issued on April 14 for an election to be held exactly four weeks later. Three major issues were on the table, each of them attracting national interest: the infant carbon tax, then just over ten months old; electoral reform; and Indigenous self-government. The first of this trio was the centrepiece. CBC News reported in early May that "environment policy seems to have grabbed centre stage."[22] But there was much more at stake than was outwardly obvious. If the Liberals lost to the phrase *axe the tax*, then carbon pricing would be "politically radioactive," said one CBC reporter. David Suzuki concurred, saying: "The carbon tax will be toxic for future politicians. No politician will raise it."[23]

Toxic. Politically radioactive. Such are not words one usually attaches to a well-designed and implemented policy, one seen internationally as a "template for the world." But that was the state of play—and the vocabulary—nearly a year after Bill Tieleman and Carole James began trumpeting "axe the tax." Academics, most economists, many New Democrats and even many conservatives understood that "axe the tax" was, as Andrew Weaver described it, "a late-nineteenth-century regressive approach."[24]

It was also an approach fundamentally rooted in smoke and mirrors. Just hours before the writ dropped, Carole James campaigned in Kamloops. Standing in front of a Chevron gas pump, she said: "Scrapping Gordon Campbell's gas tax is going to put $1.8 billion back in the pockets of average families over the next two-and-a-half years."[25] That echoed a speech to the Union of BC Municipalities eight months earlier in which she promised to keep income tax cuts in place because "British Columbians deserve a break."[26]

In a striking trans-Pacific parallel, uber-conservative Tony Abbott said almost the same thing in Australia in 2013 while working to repeal carbon pricing: "The carbon tax will go, but the carbon tax compensation will stay so that every Australian should be better off."[27] Here was the ideological opposite of democratic socialist Carole James—his black to her white—four years later and thousands of kilometres away, using almost exactly her words. That congruence on its own should be enough to demonstrate how off-base the NDP position was. And on top of everything else, both James and Abbott were quietly ignoring the revenue conundrum that "axe" would present to their respective treasuries. There was, however, one big difference between the two politicians. Abbott, as prime minister, was to prevail in repealing the tax (thereby leaving the Australian Treasury in the lurch), while James never got to put her policy into practice.

To be fair to the BC NDP, they did try to get around the revenue shell game—at least partially—by proposing to replace the lost revenue with new taxes on "big polluters," in particular oil and gas companies. It was a confused campaign position, one that the Pembina Institute described as "eliminating the foundation of the existing BC climate plan without offering an equivalent or improved replacement."[28] Pembina's calculations indicated that the NDP plan would replace the 76 per cent coverage of carbon emissions under the broad-spectrum existing tax with a "limited cap on industrial polluters" that would address, at best, just 32 per cent of emissions. For those who cared about taking action on global warming, there was little there to like.

During the campaign, James said she would agree to disagree with environmentalists. While that might have been intended to be an olive branch, it instead was perceived as pouring salt into the wound. That perception reached its apex on April 15 when Tzeporah Berman wrote a personal letter to James, the contents of which were leaked to the *Vancouver Sun*.[29] Berman pulled no punches in the first few lines: "I feel deeply betrayed by the BC NDP platform released last week and your flip-flops on green economy and climate policy. You are playing partisan politics with our children's future. Your remark that the NDP and environmentalists 'agree to disagree' could not be further from the truth." The letter concluded: "My goal remains to drive all parties into a race-to-the-top with policies to create a green economy and combat global warming. You had every opportunity to exceed BC's existing plans on green economy and climate progress. But under your leadership, the NDP, far from racing to the top, has gone into reverse." Many of those words graced the lead story of the *Sun* two days later, under the headline "Key supporter quits NDP over carbon tax."[30] It was a devastating blow that pushed James very much onto her back foot. She had been publicly called out for trying to defend the indefensible.

Why the NDP was so intent on removing the carbon tax remains a bit of a mystery. "I'm not sure what motivated them to do this," Andrew Weaver said then. But his University of Victoria political scientist colleague Jamie Lawson saw it as an attempt to "win votes in resource communities."[31] Lawson had his finger firmly on the pulse of BC politics, then as now. "The NDP must have been thinking very clearly about the electoral implications of all of this. But to me it's an open question whether the votes gained in rural BC will exceed the votes lost in urban environmental circles," he told the *Sun*. *The Globe and Mail*'s Gary Mason had a similar take: "The NDP smelled a political opportunity," he wrote, suggesting that the party was attempting to cash in on "rural anger."[32]

At least some NDP insiders shared that view. One MLA said in an off-the-record interview that "it seemed like a hot political winner," adding that an attack on the tax "might allow us to capture some constituencies in rural BC. I think there was some simplistic thinking that it would be enough to put us over the edge." That position—promoted by "the so-called rural caucus"—put many of the urban NDP members in an uncomfortable corner because "working with the team was important," says the MLA, adding that some felt obliged to sacrifice principles for the sake of solidarity. When asked if the decision came from the top, the member replies: "It wouldn't have happened if it didn't have Carole James's sign-off, because it was a fairly centralized operation within the leader's office at the time."

Today, James sees it differently, suggesting that rather than being centralized, the decision was collective. "No one faction or person drove those discussions. We really did spend a great deal of time coming to a decision," she says.[33] "We were not opposed to a price on carbon emissions." Rather, "we felt [then] that giving high-income earners a tax break was not equitable."

But in 2009, taxing only the big polluters was based on a fundamental misinterpretation of the Campbell tax. High-income earners were given no break—none. "Looking back, I feel that I—and we—didn't communicate our position well enough," James says. "We were seen as anti-climate action, and our position became caught up in the 'axe the tax' campaign rather than an alternative implementation position on carbon pricing."

Any electoral gains the New Democrats had hoped to make in the BC hinterland came to nought. In the northern half of the province, where the economy is highly dependent on exploiting natural resources, the electoral result was a draw. The NDP flipped one seat but lost another. Campbell's government retained both its share of the popular vote and its legislative majority.

Did "axe the tax" contribute to the NDP's disappointing result? Maybe, suggests UBC political scientist Kathryn Harrison. "Although Premier Campbell certainly had not run on the carbon tax, it is nonetheless striking that his acceptance speech on election night placed the carbon tax front and centre," she wrote.[34] "The premier argued forcefully that the election results 'send a message to others who may have looked at this with trepidation.'"

Years later, Campbell told me that he believes he won the election based on climate policy, estimating that some 3 or 4 per cent of the popular vote came from constituents who didn't care much about politics, but did care about the environment. Harrison puts the emphasis elsewhere, concluding: "What seems clearest is that the carbon tax had at best a minor impact on the outcome of the election, since by the spring of 2009, the issue had long since been eclipsed by the economy." Harrison's view is supported by polling data that ranked economic concerns as being front of mind—no surprise, given how hard the global recession of 2008–09 had hit the province (and the rest of Canada).

There is a critically important message to take away from BC's experience with "axe the tax": it failed. It remains vital to recognize that a conservative pro-carbon-tax government was re-elected despite a populist campaign misguidedly designed to overturn sound emissions policy. The right path won out. Gordon Campbell understood that.

So did some NDP candidates. Spencer Chandra Herbert, running in Vancouver's West End, had wanted nothing to do with "axe the tax." It would not go over well in his densely populated urban constituency, "where most of us walk or bike or bus."[35] He directed his staff to avoid any mention of "axe the tax" in his campaign literature. He also rued that the NDP's environmental initiatives were either ignored or were drowned out by "axe the tax." In his mind, it was an oversimplified smokescreen that sacrificed potentially good NDP policies on the crass altar of populism.

Carole James soon came to understand that. Just thirty days after her disappointing election result, James shuffled her shadow cabinet. She appointed a new critic on environment policy, Victoria-area MLA Rob Fleming, known for his bright green credentials. James told the *Vancouver Sun*'s Vaughn Palmer that Fleming's appointment signalled an end to the "axe the tax" campaign.[36] "The election is over and the tax is here to stay," she said. Fleming's job would be to "show people in BC that we have a very positive agenda on the environment." Palmer couldn't help himself, snidely adding: "As opposed to the mostly negative 'axe the tax' campaign." The *Victoria Times-Colonist* editorial board was similarly blunt in its appraisal of James's epiphany, writing: "Carole James has finally admitted what most British Columbians recognized long ago: The party's silly campaign to axe the carbon tax was a dud."[37]

James's *volte-face* was greeted with plaudits from climate commentators. Richard Littlemore, writing for DeSmogBlog, captured it well, headlining his piece: "Bravo! Carole James."[38] With that "one, sure gesture" that offered redemption, James reclaimed at least part of the high ground in ethical politics.

The "axe the tax" debacle had, one hopes, been dispatched permanently to the scrapyard of political missteps. There is a saying in the public relations world: "Admit you are wrong and move on." To her credit, Carole James did just that.

EPILOGUE: *Carole James announced her resignation as NDP leader on December 6, 2010, eighteen months after her reversal on the carbon tax. The announcement came as a surprise to most British Columbians, but not to NDP insiders. Yes, she'd led the party to two consecutive electoral defeats, and yes, the party had plateaued in opinion polls—its approval rating being stuck just below 50 per cent—but that was twice as high as the mere 24 per cent who'd said just two months earlier that they'd vote for the Liberals. Moreover, there was no obvious heir-apparent waiting in the wings, no fresh, inspiring Barack Obama–like figure knocking on the door. And Gordon Campbell's personal popularity in the autumn of 2010 had plummeted to less than 10 per cent following a serious flip-flop he'd made in agreeing to harmonize provincial and federal sales taxes. Indeed, with disarray in the Liberal ranks, Carole James may well have been poised to win the next election.*

"Politics in British Columbia is a rough, tough and often nasty business. As it is elsewhere. Just ask Kevin Rudd." So wrote Will McMartin in The Tyee *on December 10, 2010, four days after James announced her resignation.*[39] *McMartin was referring of course to Julia Gillard having spilled and replaced Rudd earlier that year as prime minister of Australia. Just as in the Labor Party there, BC's NDP had a bit of a history of eating its own. Being good, as James discovered, does not mean being good enough.*

But a key question remains: Did "axe the tax" contribute to James's demise? It's impossible to know. As McMartin put it, "the floodgates of dissent" became evident in the NDP in October 2010, when MLA and forest-policy guru Bob Simpson was mildly critical of her leadership. Simpson had been one of the party's MLAs who openly objected to "axe the tax." His comments opened the door to other dissenters. Some observers drew a straight line between James's hostility to the tax and the decline in internal support for her leadership. University of Victoria political scientist Denis Pilon said that her "position among many NDP activists was harmed by her aggressive opposition to Campbell's carbon tax."[40] That earlier opposition may have come back to bite her, eighteen months after she'd recanted her take on the tax. But Spencer Chandra Herbert thinks this might be overstating the importance of "axe the tax."

"I'd say it was more of a general management leadership issue," he said. "We needed a reboot."

In December 2010, a disapproving public statement released by veteran MLA Jenny Kwan proved to be the last straw. James saw the political writing on the wall and announced she would step down.

There is irony here. Insiders suggest that Kwan's statement may have borne the fingerprints of "axe the tax" promoter Bill Tieleman, a suggestion he would neither confirm nor deny, according to Richard Littlemore.[41] On December 8, 2010, James told the Vancouver Province that Bill Tieleman—who had stood shoulder to shoulder with her in the "axe the tax" effort—had indeed been "actively working to remove her."[42]

With apologies to Thomas Hobbes, yes, politics in BC can be nasty, brutish—and unforgiving.

Just like in Australia.

CHAPTER 8

Going Wobbly on Climate Action

The Christy Clark Years in British Columbia, 2011–17

Of the many mistakes that she made, the most strategically significant was in abdicating her responsibility to lead the shift to a new low-carbon economy.
MARTYN BROWN, June 2017

NATURAL GAS: BRITISH COLUMBIA HAS LOTS OF IT, BOTH IN CONventional easy-to-tap pools and as unconventional shale-hosted pockets that can only be released by the relatively young practice of fracking. And any world map will show you that the province is, quite obviously, Canada's gateway to the Pacific. Because of that portal, British Columbia was bisected in 1953 by the Trans Mountain Pipeline, a major conduit to the world market for landlocked Alberta oil. That pipe has carried both crude and refined hydrocarbons from the continental interior to tidewater near Vancouver for more than seven decades.

The combination of natural gas and that pipeline—plus further bitumen extraction in northern Alberta—add up to a hydrocarbon-based election strategy, one that promises royalties and jobs. At least that's what Premier Christy Clark thought in the provincial election campaign of 2017. Instead, it was a failure. While that failure was a disaster for the BC Liberals, a bigger disaster had preceded it: over six years, British Columbia had, increment by increment, stepped back from its internationally applauded leadership on the climate-action file.

Those six years had a rocky start. On July 23, 2009, some ten weeks after the Liberals handily won a provincial election, Gordon Campbell announced that British Columbia would harmonize its provincial 7 per cent sales tax with the federal 5 per cent goods and services tax. The merger was to yield a single 12 per cent tax (the HST) on almost all goods and services, including some items—restaurant meals, haircuts and school supplies—that had not previously been subject to the provincial sales tax. For most consumers, it was received as generally grim news.[1] Vaughn Palmer, political affairs columnist in the *Sun*, zeroed in on the impact on haircuts and buying a daily newspaper, asking: "Is nothing sacred?"[2]

Harmonization did, however, offer a particular benefit to British Columbia: the federal government was offering a one-time $1.6-billion transition payment, which Campbell intended to use to buttress health and social services as the province worked to recover from the 2008–09 global recession. But in the election campaign a few months earlier, Campbell and many of his cabinet ministers had repeatedly said that a harmonized sales tax was far from their minds. It was a ruse that "worked to election day," Palmer suggested; behind the scenes, discussions about bringing in the tax had been quietly held. When Campbell announced that his government would move ahead with the HST, many British Columbians felt they had been duped by a dishonest party that said one thing during the election campaign, only to do the opposite mere weeks later.

Bill Tieleman—the populist political cow-pie-disturber—did his part to ramp up the growing resentment on Facebook, as he did against the carbon tax. His anti-HST rationale was both simple and similar: "What I saw as the bottom line basically is: consumption taxes are regressive taxes and so it doesn't matter whether it's the HST or a gas tax or a sales tax—they are all regressive taxes," he says. But this time he was right. Unlike the provisions in the carbon tax act that rendered it more progressive than regressive, no direct fiscal rebates were to be provided to BC consumers for the broadened tax base of the HST.

In the end, Gordon Campbell could not withstand the withering and unrelenting criticism over the waning months of 2009 and through most of 2010. By October 2010 his popularity had fallen into single-digit territory. Humorist Steve Burgess quipped that Campbell's popularity, at 9 per cent, was "running slightly ahead of belief in the Loch Ness Monster, and dead even against cherry vanilla ice cream."[3] While other scandals and issues undoubtedly contributed to the premier's unpopularity, it was the HST affair that was central to his fate. "It's time for a new person to lead," he said at a press conference on November 3, 2010, his voice cracking.[4] "It's not always popular to do what you believe in your heart is right," he said, singling out "leading the way on climate action" as one accomplishment of which he was particularly proud.

The Campbell era in BC politics was over. The conservative who established a carbon pricing scheme that was internationally recognized as "a template for the world" was leaving politics. Those promoting serious climate action immediately began to wonder what would be next.

The global context at the time was simultaneously troubling and exciting. NASA was about to announce that 2010 was tied for the warmest year on Earth since thermometers were invented. BC had just lost a massive piece of its forest stock to the mountain pine beetle, owing primarily to warming. The Arctic Ocean was losing summertime sea-ice cover and progressively becoming a dark, heat-absorbing panel instead of a sunlight-reflecting mirror. Sea level was rising at a rate nearly twice that of thirty years prior, as the ocean continued to warm and expand. Alpine glaciers around the world continued to melt, pouring water into the seas. And the upper ocean was slowly becoming acidified as carbon dioxide from fossil fuel combustion dissolved into seawater, making carbonic acid that threatened coral reefs with ultimate extinction. In the face of such tightly documented phenomena, it would have been easy to lose hope.

But it wasn't all doom and gloom. Administrations around the world were acting on the climate challenge—slowly perhaps, but acting nevertheless. Many were promoting cautious excitement, generating a sense that something could be done, that something *was* being done. In December 2010 alone, Japan announced an escalating carbon tax, designed to curb fossil fuel consumption; California's Air Resources Board voted to adopt the large-emitter cap-and-trade provision laid out in Arnold Schwarzenegger's Clean Energy Act of 2006; and Chris Huhne, minister of energy in the UK, set out a four-point program that included a national carbon tax tied to reductions in income and employment taxes. One month, three disparate major economies many thousands of kilometres apart, three positive steps.

Then, as now, taking such actions—curbing carbon pollution and slowing global warming while putting money in people's pockets—made obvious sense. But additional benefits were becoming obvious. California's renewable energy sector was growing more rapidly than any other segment of its economy, providing tens of thousands of new, well-paying jobs while broadening the tax base. In

BC, Campbell's carbon tax was stimulating capital investment and innovation that was helping to drive the GDP associated with the green economy from some $15 billion then to what was projected to be twice that by 2020, accounting for as much as 15 per cent of provincial output. Some 225,000 direct and indirect green jobs were forecast to be in place by 2020, up 50,000 or so relative to 2010. Like other savvy regions around the world, BC was in the thick of recognizing opportunity, of recognizing what the global renewable energy revolution was about to generate.

By January 2011 it was abundantly clear that much was at stake in maintaining the green momentum established under Gordon Campbell's leadership. It was no time to "go wobbly" on the climate file, as Margaret Thatcher would have put it. And yes, that's the same Margaret Thatcher, a hardcore conservative who—in her landmark speech to the General Assembly of the United Nations in November 1989—forcefully exhorted the nations of the world to face head-on the challenge of reducing CO_2 emissions. "We can't just do nothing," she said, emphasizing the need for immediacy, and noting that industrial countries bore a special responsibility to act.[5] Weaning economies off fossil fuels through establishment of incentives and policies was a—maybe *the*—critical directive she offered that day to the 192 international delegations in the cavernous General Assembly Hall.

But that message didn't reach all leaders everywhere. As political and economic winds buffeted governments everywhere in the years following, it lost staying power. Enter another market-oriented conservative, two decades later: Christina Joan Clark.

Christy Clark was well known to British Columbians, having served as a member of the legislature from 1996 to 2005, including stints as minister of education (2001–04), minister of children and family development (2004), and four years as deputy premier

(2001–05), after which she temporarily left politics. Clark burnished her gregarious public persona as a talk-radio host in Vancouver from 2007 to 2010 before deciding, in the wake of Campbell's resignation, to run for the BC Liberal leadership. She was the fifth candidate to announce her candidacy, the previous four being sitting cabinet ministers and key members of Campbell's inner circle. Despite her prior experience, Clark cast herself as a political outsider who could bring change to the party. She narrowly won the leadership in late February 2011, and was sworn in as premier a fortnight later.

Christy Clark was highly personable, charismatic, articulate, quick-witted and socially liberal. While she was also a staunch proponent of what the conservative element in BC called "free enterprise," she was no darling of hardcore right-wingers, many of whom held a grudge for the four years she spent as a talk-show host critiquing uber-conservative Prime Minister Stephen Harper. Others saw her as policy-light. One columnist wrote during the leadership campaign that "with Christy Clark, you certainly get lots of sizzle. Substance is a different question."[6] Substance or not, she was soon to demonstrate she had impressive political smarts. But did those smarts include the climate challenge? Would she continue to drive the burgeoning green-energy economy? More specifically, would she stay on the straight-ahead carbon-pricing rail line that Gordon Campbell had laid down? Or would she wobble and shunt BC's carbon tax onto a siding—a move that would have led Margaret Thatcher, had she been aware, to raise an eyebrow.

Clark's leadership-campaign platform was free of obvious political risks, promising to put "families first," create jobs, balance the budget and get the HST initiative over and done with. It also lacked vision. *Families first* is crisp, clean and appealing on a personal level, a two-word phrase any advertising agency would salute. But in early 2011, another two-word phrase—one that demanded a visionary policy response—could not be found in any of her position statements: *climate change*.

That absence heralded two unwelcome policy directions: a massive increase in the production and export of natural gas, and freezing BC's carbon tax. Within two years, both were to become political hot potatoes. Another lurked in the background: twinning pipeline capacity across BC to Pacific tidewater to facilitate the expansion of bitumen mining in Alberta's north. All three issues were underlaid by one common theme: greenhouse gas emissions and their climatic implications. It was that theme that eventually was to play a key role in ending Clark's tenure as premier.

British Columbia's fixed-date election law required that a vote be held on the second Tuesday in May 2013. By the time the writ was dropped a month earlier, Premier Clark and the BC Liberals had become highly unpopular. Poll after poll had the party running well behind the left-wing NDP, the result of residual opprobrium over the HST issue, the carving-off of hard-right support to a newly led and freshly invigorated BC Conservative Party, not to mention a limited but festering suite of background scandals.[7] Moreover, the Liberal caucus was far from united. There had been "two years of internal warfare," Premier Clark says—mostly, she claims, by MLAs who felt disregarded by Campbell and left out of cabinet.[8] "Every single day the House was in session, I would stand up in front of caucus, because we had a daily meeting, and I would more or less beg them not to do anything dumb or destructive when we got into the House." It was not the best of times to seek re-election.

But campaign she did, and she was good at it. As the clear underdog, Clark went on the attack, making the economy the centrepoint. She argued relentlessly that only continued economic growth could support the social programs British Columbians had come to expect, as well as health, education and debt reduction. She hammered the

NDP, claiming that they would tax-and-spend the province into penury. It was old-style scare politics. While she used that shopworn approach to effect, she was more eager to focus on a two-pronged hydrocarbon strategy she saw as fresh: pipe vast volumes of diluted bitumen from Alberta's tar sands to export facilities at the BC coast—an effort strongly supported by Canada's oil industry *and* the Government of Alberta—while also producing and exporting large volumes of LNG (liquefied natural gas), using feedstock methane to be won from the ground by horizontal drilling and fracking deep-seated shale formations in BC's northeast corner.

For fifty years, mostly conventional oil had flowed from Edmonton to the Vancouver suburb of Burnaby through the Trans Mountain Pipeline. By 2013, an increasing proportion of the flow consisted of diluted bitumen, or "dilbit," a viscous tarry amalgam of hydrocarbons mined from the Athabasca tar sands and thinned by adding low-viscosity compounds like naphtha and propane. There was a waiting refinery market for the product. The then-owner of the pipeline, Texas-based Kinder Morgan, had for the previous three years mused about twinning the Trans Mountain and tripling its liquid hydrocarbon transport capacity to 141,000 cubic metres per day.[9] It was a sharply polarizing initiative that had been bubbling in the background, causing consternation among environmentalists, the tourism industry and others with a stake in "Super, Natural British Columbia."

Twinning the line would lead to a sevenfold increase in tanker traffic seaward from the Port of Vancouver through the narrow passages of the Gulf Islands of British Columbia toward the open Pacific. It was a festering issue that presented political peril for the provincial government, despite offering what many in the resource sector touted as a risk-offsetting economic boon. Christy Clark and her team implicitly recognized the incendiary nature of the issue and worked to defuse it.

Terry Lake was environment minister then, having been appointed in March 2011 to Clark's first cabinet. A veterinarian early in his career, Lake had first been elected in 2009. "I ran for the Campbell Liberals because of the carbon tax and their policy on climate change," he says, noting that when he became a minister it was "like a dream come true, because things like climate change were very important to me."[10] But in Lake's view, dealing with the climate challenge did not exclude allowing pipeline expansion. Asked in 2022 about the Kinder Morgan proposal, he said, "Well, obviously, I would prefer that we didn't have the need for it, but I also understood the political and economic realities."

Under Lake's watch, the Clark government had opposed a competing proposal, the Northern Gateway Pipeline, designed to carry dilbit from Alberta to tidewater across the wilds of northern British Columbia. Looking at the proposed Northern Gateway route, Lake saw the obvious: "Heavy oil was going to be traversing almost inaccessible areas of the province, with vulnerable waterways, and any leak would go unrepaired for ages and the impact would be enormous." In contrast, the southern route made more sense to him: There was an existing pipeline on the route—the Trans Mountain—that had been there for sixty years. He and Clark saw exporting dilbit through a newly twinned southern line as a pragmatic acknowledgement that "we're not going to get off fossil fuels tomorrow." But defusing the growing opposition to the pipeline remained a challenge in mid-2012, with the next election just ten months hence.

At a packed press conference on July 23, 2012, Terry Lake and Aboriginal Affairs Minister Mary Polak stood shoulder to shoulder and announced their plan to finesse future twinning of the pipeline.

Five key conditions would have to be addressed before any new pipe would carry heavy oil across British Columbia: passing environmental review; putting in place "world-leading" oil-spill prevention on both the ocean and land, as well as response and recovery systems; respecting treaty rights and offering opportunities to First Nations communities; and delivering to the province a fair share of fiscal and economic benefits.[11]

Lake recalls the five-conditions requirement as an accommodation, a way to find middle ground. The conditions were "high standards," Lake insists, and they were needed to "justify the political capital for our position. I thought it was very elegant."

It was a smart way to buy political time, and it worked. It took the issue "off the table for a little bit," says Lake. As a premier seeking to stay in office, Clark could now claim she was on the side of citizens protesting the likelihood of many more oil tankers plying the Strait of Georgia's choppy waters.

She took a different tack on LNG, however.

Fracking was then the new technological darling of the fossil fuel industry. It requires shale to be split and cracked open ("fracked") underground by very-high-pressure injection of fluids bearing chemicals and sand through multiple drill holes. Trapped methane, the primary component of natural gas, is released from interstices within the shale strata and collected. It's then treated to remove impurities like carbon dioxide, and pumped to markets—or, increasingly, to liquefaction plants where it's subjected to intense refrigeration and cooled to at least -160° C, the temperature below which gaseous methane exists as a liquid. Christy Clark saw this technology as an economic engine.

In the early 2000s, exported LNG on the world market was commanding between four and five US dollars per million metric British thermal units (mmBtu).[12] The price had been relatively stable for several years. But in the late 2000s, Asia's appetite for LNG rose dramatically. By June 2012 the tariff had more than tripled, peaking at over US$19 that month. The aroma of a gold rush was in the air—there was a lot of money to be made, and someone was going to make it. Why not British Columbia? Premier Clark thought at the time: "We realized we could access vast resources of natural gas ... and have an opportunity to export it." Clark saw the opportunity not only as a pot of gold but as a pathway to success in the 2013 provincial election. It was a prospect that in her mind fit hand-in-glove with job (and wealth) creation, something she'd promised in the leadership campaign of 2011. All she had to do was sell the idea to investors and the BC public.

She was a good saleswoman. Her task was to convince citizens to drill thousands of deep holes into sedimentary formations across northeastern BC, pump vast volumes of chemical-laden water into those holes at high pressure, and recover the methane thus released from the split rock, along with high volumes of contaminated water that had to be carefully disposed of, and then pipe the gas long distances across the mountains and rivers of the BC landscape to newly built plants on the shores of our largely pristine Pacific fjords, where it was to be liquefied and exported to Asia via large ships that would have to navigate without incident the ins and outs of rugged and treacherous coastal topography. Easy, no?

No. Not easy.

Establishment of a new LNG industry in BC faced several major hurdles. Putting aside the obvious negative environmental issues associated with fracking itself—water consumption, land-use demand, contaminated-fluids disposal, fugitive methane emissions—BC's shale-gas fields were far inland. Long-distance gas transport pipelines would have to be constructed, running westward across the northern half of the province from near Alberta to termini just south of the Alaskan Panhandle. Permission from a number of First Nations, across whose traditional territories the pipelines would transit, would be obligatory under BC's existing legislation.

LNG production is energy-hungry, requiring vast amounts of electricity to run compressors that drive the refrigeration systems. Because BC's existing electrical grid could not supply the power needed to run the compressors, natural gas–driven turbines, a.k.a. direct drives, would be the power supply of choice for the refrigeration plants. Natural gas turbines emit CO_2, and lots of it. But the province's climate-action plan, produced during the Gordon Campbell years, mandated *reductions* of carbon dioxide emissions. Proposed LNG plants that were to operate via direct drives promised to *increase* BC's aggregate emissions by very large numbers, on the order of millions of tonnes of carbon dioxide annually.

The capital investment required to build the proposed plethora of pipelines, plants and docks was on the order of tens of billions of dollars. BC would be competing with other jurisdictions also seeking investment on similar scales, and that were at least equally keen to cash in on the LNG bonanza. How could or would we attract investors when our proposed operations looked to have higher production costs compared with countries like Qatar or Australia, where rich natural gas fields sat in coastal settings, and where emissions reductions were afterthoughts at best?

The economic foundation of the proposed industry was strictly tied to the export price of LNG. There was no guarantee that it would remain at the stratospheric levels seen in much of 2012; indeed, the price of any fossil fuel had roller-coasted wildly over the preceding four decades. Why should this boom be any different, and not end in a bust?

Finally, and perhaps most important, was the moral dimension, one that drew a straight line to the intense obligation to reduce emissions. How could BC meet that obligation? Christy Clark's position on this was simple, claiming that by using LNG to replace coal in electricity production, emissions would be cut by a third. In her mind, if China, for example, was to use BC's LNG to replace coal-fired power production, the global atmosphere would be better off. "Why didn't Gordon [Campbell] do it?" she muses. "Gordon had a different view fundamentally of the role British Columbia would play in solving the world's problem. He really was quite strict about his view, which was [that] British Columbia is going to lower its own emissions, period."

What Campbell was promoting was direct environmental responsibility. What Premier Clark envisaged was an indirect rationalization.

Those choices were not the same.

Not surprisingly, Clark met no opposition from her own caucus, which readily green-lit what they saw collectively as a once-in-a-lifetime opportunity. She says natural gas did not fit into Campbell's view, but that her caucus welcomed it "because some of them really just didn't care about the climate agenda. But for me, whether or not they cared about the climate agenda, I did. I thought it fit. And if they were going to support me on it, good. We were going to go ahead and get it done." She was comfortable with the mashup: burning BC's LNG elsewhere would reduce net emissions by substituting for nastier coal, even if that meant BC's emissions-reductions targets were to be discarded. Terry Lake supported Clark's position. "I always was a believer that natural gas was a transition fuel. I'm a pragmatist, I'm a realist," he says. "If it displaced coal power generation in Asia, the net [emissions] benefit would be there. And guess what? We'd make a lot of money, employ a lot of people."

Cashing in on a methane bounty weighed heavier than concern about net global warming. Clark's government at the time was touting the fiscal potential of LNG in terms that bordered on the fantastic. The Budget and Fiscal Plan released in February 2013 suggested that provincial revenues from LNG exploitation—royalties, general corporate, sales and income taxes, a proposed specific BC LNG tax—would run from a minimum of $79 billion to a maximum $185 billion over twenty years.[13] On February 12, 2013, three months before the pending election, Clark announced the creation of the British Columbia Prosperity Fund, a new governmental wallet to be stuffed with "more than $100 billion" of revenue from the anticipated LNG industry.[14] The fund would be used to ensure that "communities, First Nations and all British Columbians benefit" from the development of the resource.

There was more. Construction of five proposed plants could generate "more than 39,000 average annual full-time jobs" over nine years, and "more than 75,000 permanent annual full-time jobs could be required to support the industry when the plants are fully operational." It was heady stuff that promised wealth while quietly shoving emissions behind a less-than-green curtain. And promising to deliver wealth is rarely a poor election strategy.

Premier Clark had one more electoral ace up her sleeve. On April 3, the *Vancouver Sun* reported that her government, if re-elected, intended to freeze BC's carbon tax for five years, effective July 1.[15] The move would not only "help make life more affordable" but "level the playing field for local industries who are competing against jurisdictions that don't have a carbon tax," she said that day, adding: "We believe in lower taxes."[16] But some industrial sectors, in particular the rapidly growing cleantech industry in the province, saw it differently. Jonathan Rhone, then head of the BC Cleantech CEO Alliance, pointed out that the market for clean technologies and climate technologies was burgeoning. "There was a global demand for what we were doing," he says, and at the moment when Clark froze the carbon tax, "we forfeited our leadership."[17] Then the province "doubled down on LNG, and made up this falsehood that if you export LNG to China, it will somehow reduce the number of coal plants that they build. That intellectual dishonesty of the government was really hard for those of us who were in the clean energy sector." But Rhone points out a positive in Clark's actions that month. To her credit, he suggests, she did not go beyond the freeze. "She did not dismantle all of the climate-action infrastructure, and she was under pressure to do so. She could have done great harm. And she didn't."

The sideways shift wasn't universally supported within her own party. At least two key cabinet ministers saw it as a misstep. In Terry Lake's view, it was a regressive, albeit politically astute, move. Mary Polak, then minister of transportation and infrastructure, is less charitable while accepting part of the blame. "One of the most foolish things we did politically was freeze the carbon tax," she says.[18] People had become used to it, she says. "The annual increases happened. They were small. People were getting money back on their taxes. All was right with the world." But "when you freeze something, you set yourself up for the big reveal. When is it going to be raised? That will always be politically challenging."

Cabinet uneasiness aside, Clark saw linking the freeze to affordability as a political winner. But what the average citizen might have missed is that the premier was playing a disingenuous shell game. The carbon tax *was being used to reduce other taxes*, on both corporate profits and personal income, *and* it provided direct cash subventions to families and individuals at the low end of the income spectrum. Those reductions and supports had been, and were still, substantial. In that light, any bigger-picture analysis would have revealed the vacuity of the carbon-tax-freeze justification. But how many voters, or reporters for that matter, even noticed they were being taken for patsies? Revenue neutrality—putting money back into both corporate and individual taxpayers' pockets—should have been front and

CHAPTER 8

centre in any discussion of freezing the carbon tax, but in the heat of the 2013 campaign, that obligation appeared to have been lost.

Years later, with unrecognized irony, Clark recalls how important the revenue-neutrality provision had been in BC's carbon tax. It "was vital to the success of the plan," she says, and it's why the tax "was recognized by the World Bank and the United Nations." Without revenue neutrality, "a carbon tax just becomes a villain" and can become a "tipping point into economic slowdown." Most economists would agree that is correct. Well-designed neutrality keeps the economic see-saw from tipping, as Gordon Campbell fully recognized.

But in the 2013 campaign, Clark portrayed the carbon tax differently, implying that it needed to be frozen because it was imposing economic hardship. "Now we're at a point where people are finding it really difficult to afford to live in British Columbia," she told the CBC then, citing the unstable world economy.[19] "Freezing the carbon tax for five years is part of my commitment." In justifying that position, she was employing deception by omission, conveniently failing to mention the tax relief and direct help for low-income citizens provided from carbon tax revenue. When asked how her proposal differed from the platform of the NDP, she proclaimed they "would raise the carbon tax. We will freeze it."[20] Simple. Crisp. And unfettered by any attention to detail or concern about demoting BC's climate leadership.

Five weeks later, Clark's Liberals won a clear majority, claiming forty-nine of eighty-five seats. It was the fourth straight majority for the party, and the second since taking climate action became a formative direction in provincial politics. The election result confounded the pollsters, almost all of whom had predicted victory for the left wing. Media commentators were gobsmacked. Veteran legislative reporter Michael Smyth, writing in the *Province*, captured the surprise well: "Go ahead and call it what it is: the most amazing political comeback British Columbia has ever seen."[21] Christy Clark, the obvious underdog coming into the race, "played the strongest card in her hand over and over again: the economy," while the New Democrats ran a "timid and cautious campaign" that failed to counter Clark's consistent message that the NDP was "a threat to your job." Asking what would happen next, Smyth zeroed in on Clark's LNG vision "that could generate so much wealth for the province that she could eliminate the province's debt in just fifteen years." But, wondered Smyth, could she deliver?

Andrew Weaver had an answer. It was, in a sense, maybe/but, with emphasis on the latter. Weaver, one of Canada's most celebrated climate scientists, ran for office in the election and became the first-ever Green Party member to sit in the legislature. On July 3, he delivered his maiden speech, a passionate plea to respect future generations, while offering a warning.[22] "I believe that, ultimately, we all want the same thing: a future that ensures that the needs of both present and future generations are met, a future that provides our children with opportunities for innovation and prosperity. But we must be cautious of striding forward, heedless of the obstacles before us ... We must not rest our future on one desired outcome," he declared. Weaver was acutely cognizant of the profound disconnect in Clark's energy policy, saying: "To attempt to lay all our eggs in this one basket on the hope, the desperate hope, that we will be the one exception, the one jurisdiction that will defy history and economics, is to base our very future on unstable ground. Yet even if we defy the odds, coast through the boom-and-bust cycles ... we cannot achieve our provincial carbon targets while we double down on LNG."

Those words captured directly the moral quagmire facing Premier Clark and her team. Looking back years later, Clark would have none of it. "I [didn't] accept his premise," she states. "It [was] wrong. Our jobs plan had eight core objectives in it, from eight different core export sectors of our economy." LNG was a brand-new industry, she says, "so we were building it from nothing. We were building the tax regime, the environmental regime." She's firm in her belief that "we were *not* putting all our eggs in one basket. That's a myth that that the opposition endeavoured to create."

But was it a myth? In the few years after the 2013 election, only one of the premier's eight core objectives—liquefied natural gas—was trotted out, day after day, month after month, as the primary economic stimulus that was going to propel British Columbia firmly toward debt-free paradise. That stimulus, however, largely ignored the stark choice that, in reality, faced the province. Weaver again framed it well, echoing Gordon Campbell: "We can offer future generations a path that pits the economy against the environment for narrow-sighted economic gain, or we can be the generation that leaves a legacy that rejects this outdated thinking, one that invests in a strong and enduring economy that is founded on a culture of unwavering environmental responsibility."

In the end, environmental responsibility was to play second fiddle. Christy Clark's government worked tirelessly to establish a vibrant LNG industry in the province, even as science increasingly showed that raw methane emissions during natural gas production were rising steadily, alongside escalating fossil-fuel combustion rates. That combination was ratcheting up global warming, even as LNG fever in British Columbia continued to sweep across the political and industrial landscape.

Tax revisions, subsidies, regulatory changes—all were up for discussion in an effort to attract capital. The fossil fuel industry repeatedly made its case, implying—if not stating outright—that deep-pocketed investors could readily invest somewhere other than in British Columbia, or Canada, for that matter. Dave Collyer, president of the Canadian Association of Petroleum Producers and a former

Royal Dutch Shell executive, made that point on the national landscape. "These new market forays face stiff competition from other exporters, particularly Australia. Timely development with an efficient regulatory process and a competitive fiscal regime are key to successfully advancing Canada's interests," Collyer said[23] in 2013. "Competitive fiscal regime" was the lever—some might say cudgel—that multinationals were wielding so effectively against a government so keen to put an LNG win on the table.

Alberta, too, was in the picture, but not as a centre for LNG production. Rather, interest there—as much as it was, given the context that oil, and not gas, was king—was focused on sending Alberta gas westward to liquefaction plants on the coastline. Christy Clark saw that as a win-win for BC, provided that she could lure to the province the investment she needed. "Natural gas was always really an afterthought for [Alberta]," she observes, "which was part of the economic opportunity for us." She wanted to capture LNG business for BC, and then capture added value from Alberta's gas that would certainly be processed at tidewater in BC. "We started to structure our deals with companies like Petronas and Shell with an eye to making it more competitive for [those companies] to extract their gas in British Columbia," she says, "because we wanted the whole cake. We wanted the cake and to eat it too."

In structuring deals to get that win, the premier and her team ceded a lot of potential revenue to the multinationals. Clark's first full post-election budget set the stage for such future backtracking. In presenting the budget on February 18, 2014, Minister of Finance Mike de Jong reiterated the government's golden vision, saying: "Any conversation about the future of BC must take into account the huge potential represented by LNG."[24] Realizing it, he offered, would "require a tax regime to help ensure that the people who own the resource—the people of British Columbia—derive their fair share of benefits." To that end, he laid out the future tax structure that would yield the "fair share": a two-tier tax, "levied on net income from liquefaction of natural gas at LNG facilities in British Columbia. We are proposing a tier-one tax rate of one-and-a-half per cent, that would apply at the commencement of production, and a tier-two rate of up to seven per cent that would apply once capital investment costs in the LNG facility have been deducted."

Up to 7 per cent. While that number struck most commentators as reasonable, de Jong went on to give himself plenty of room to revise it, in code thinly disguised to foreshadow future retreat: "The rates will be finalized in the legislation we intend to introduce in the fall and will ultimately be determined by our ongoing analysis of global economic and market conditions, with a view to ensuring that BC remains competitive for investors, and British Columbians derive a fair return for the sale of the resource they own." Translation: seven is a vulnerable number.

Eight months later, that vulnerability was fully exposed. The long-awaited LNG legislation cut seven in half. For the twenty years following full recovery of capital costs, the new tax rates were to see LNG producers pay just 3.5 per cent on net returns from the sale of methane fracked from shale. While the Opposition saw the backtrack as kowtowing to pressure from multinational corporations, Minister de Jong took a different tack. "We have been aware, from the outset, that we are being analyzed on a comparative basis with other jurisdictions," he told reporters. Based on his analysis, "we represent the single most cost-competitive jurisdiction within which [LNG producers] can establish themselves." But there can be little doubt that industry had successfully applied the screws. "BC was pressured to rejig the proposed tax rate," wrote a team of Toronto-based tax and trade lawyers, in one comprehensive analysis.[25]

Julie Gordon of Reuters reported that Malaysia's Petronas—a major potential investor in BC's LNG—had warned earlier in October that "it could delay its $11-billion project by more than a decade unless a favourable tax deal was reached."²⁶ Gordon noted that LNG Canada, a consortium led by Royal Dutch Shell, was somewhat more positive, describing the legislation as an "important step forward in providing the certainty that companies need." But the fiscal writing was clear. The *Vancouver Sun*'s Vaughn Palmer saw what was coming: "When it comes to driving a hard bargain, it strikes me that it is the Liberals who can't afford to walk away from the table on LNG, not the multinational oil and gas companies," he wrote. It was the multinationals and not the BC government who carried the big stick.²⁷

Years later, neither Christy Clark nor former Environment Minister Mary Polak look at the revised tax regime of late 2014 as a giveaway.²⁸ In their view, it was the reality of the global market. The corporate sector had smartly pushed administrations around the world for minimum resource rents, playing gas-producing regimes like Australia, Louisiana and Qatar against a government in British Columbia that was so publicly keen to establish one more resource industry. Slashing the "fair share" of February 2014 in half a mere eight months later—and setting the low rate for twenty years, no less—smacked of a government overly eager to claim a stake in the global LNG marketplace. It was akin to a grade-school student desperately waving their arm in the classroom, imploring the teacher to "pick me, pick me!"

To this day, Clark sees it differently. "The LNG tax was something that we conceived back when the price of natural gas was through the roof. It was booming. It looked to us and the finance folks in government like there was a bigger slice for government to be had there. But by the time we ended up actually making the deals, the price was a lot lower and that was really what drove it," she says. But price data refute that view. Yes, the tariff for LNG on the Asian market in the ten years before 2014 had been volatile, ranging from about US$5 per million metric British thermal units (mmBtu) in late 2003 to high but short-lived peaks almost four times higher through much of 2011–12.[29] But the price was as high again in 2014. In February of that year, when BC proposed a 7 per cent maximum LNG tax, the gas was commanding about US$18 per mmBtu; it was only marginally lower eight months later, at around $16, when Mike de Jong took an axe to the rate.

Since the data suggest that it wasn't price that fundamentally drove the finance minister to halve the maximum tax, what did? Could it have been the positive political optics of selling a win—of creating a new, intensive resource-extraction industry and its promise of a fiscal windfall—in combination with corporate pressure along the lines of "We'll walk if you don't sweeten the pot"? Mary Polak insists there was another issue driving the decision: the province dug in its heels on the carbon tax, which, while frozen at $30 per tonne of CO_2 emitted, remained in place. "We didn't feel we could tell the public, 'You're going to have to pay it and these guys aren't.' And we kept to that. Did that mean we had to trade off some other things? Yep." In Polak's mind, it was a fair trade: hold the line on a continuing emissions tax but yield on the tax rate for net gas sales. In the fall of 2014 the province yielded. The corporate giants got what they sought.

While the LNG negotiations were continuing, British Columbia's progressive emissions-reductions agenda sat on the sidelines, more ignored than forgotten. But the prospect of the province adding millions more tonnes per year of carbon dioxide to the global atmosphere—when multiple direct-drive LNG compressors were likely to be turned on a few years hence—was not to go away. It was a political and environmental downside, and Christy Clark was sensitive to its implications.

By 2015, as the Paris Agreement was inching toward global acceptance, Clark realized that "we needed to take another step in climate action." The economy had recovered. "Now we had the luxury of being able to pay more attention to some of the other issues that sometimes don't get the priority that they should," she reflects, "to redouble our efforts about what we're going to do with climate." It was a tacit admission that economic concerns had been trumping environmental responsibility. Clean Energy Canada's Merran Smith was frequently in contact with the premier's office. Her view was that much more than economic issues drove Clark's thinking: "She was going to lose social licence for LNG development if she didn't demonstrate action on climate."[30]

To redress that imbalance, the premier announced her next step, on May 12, 2015.[31] She created an advisory team that promised to build on the province's "climate successes." It was an attempt to insulate the government from current and future opposition to its hydrocarbon strategy. That context was captured well by Terry Lake in 2022: "We talk about shields and swords in politics, and I think the formation of that group was a bit of a shield."

Clark boasted that the nineteen-member BC Climate Leadership Team, which included this author, "brings together leaders from the business, academic and environmental communities, as well as First Nations and local governments, to consider the best actions to keep us on track toward meeting our greenhouse gas reduction targets."[32]

The words in that press release, like *consider*, had been carefully selected. Spoiler alert: they harboured no suggestion that the government would be obliged to heed any of the recommendations to come.

The Climate Leadership Team (CLT) met repeatedly for the next five months with captains of industry, civic and provincial politicians, First Nations leaders, environmental experts and economic modellers. Merran Smith was a key member and remembers those meetings as "sandwiches on large white buns, windowless meeting rooms, and endless PowerPoint." Anodyne sandwiches aside, it was a heady, enervating time. Here was a comprehensive cross-section of British Columbian society—a mini citizens assembly—tasked with laying out significant actions that, if applied, would foster ongoing reductions in emissions *while simultaneously balancing economic—and to some extent, political—imperatives*. Added to the list were the challenges of mitigating negative impacts on vulnerable populations and maintaining the province's reputation for progressive climate policies. The team faced a hard, imminent deadline: report by October 31.

It was not easy. Early on, the team accepted the reality that LNG was likely to be exported from British Columbia at a scale yet to be determined. That reality included dealing with the toughest nut to crack: How to meet the province's mandated CO_2 reductions targets in the face of such a looming emissions juggernaut? Hard questions were asked and addressed; potential actions were modelled. Could CO_2-spewing direct drives be supplanted by electric drives in LNG plants, and if so, to what extent? What should be BC Hydro's role? Could other sectors of the economy cut their emissions, vigorously if necessary, to compensate for a growing, emitting LNG industry? How could progressive increases in BC's carbon tax be applied to LNG production and the other sectors: should there be exemptions? Temporary concessions? Other incentives? And what about the "other" natural gas problem, wherein fracking and pipeline operations release fugitive methane into the atmosphere?

More broadly, the global context was firmly considered: How would increases in our vaunted carbon tax end up impacting energy-intensive trade-exposed (EITE) export industries? Could we design a lowest-possible-harm or no-harm-at-all package of EITE actions that wouldn't set back agriculture or mining or forestry or manufacturing?

And with a nod toward Gordon Campbell and his revenue-neutrality principle, could the impact of increases in the carbon tax across multiple economic sectors, as well as individual citizens, be muted by lowering the tax hit elsewhere—while at the same time spurring society to reduce its aggregate carbon pollution?

All the CLT had to do was map out a feasible plan that would allow British Columbia to have its cake and eat it too. It was, in the end, a challenge that all but one of the members of the team felt was met by the team's report.[33] Feasible? Check. Economically sound? Check. Cognizant and accommodating of multilateral trade competitiveness? Check. Environmentally responsible, under the circumstances? Check. Endorsed by the Christy Clark government? Well, no, at least not as hoped.

On Halloween morning, the team delivered its recommendations. The report offered thirty-two threads, each in the form of a recommendation, woven together to yield a canvas on which was painted the broad sweep of a lower-emissions future. Pulling any of those recommendations out of the mix might cause the canvas at best to fray and at worst to unravel. It was, in our view, a package not to be tampered with. Christy Clark's government didn't see it that way.

Twenty-seven days after its submission, the province publicly released our report on a Friday afternoon. Minister of the Environment Mary Polak described it as thoughtful and "full of innovative approaches that respect the need to protect industry competitiveness and family affordability." In thanking the CLT, Premier Clark noted that it had been "tasked with an important and challenging job, to consider a broad range of actions across our province's industrial, transportation and building sectors, helping us move our climate agenda forward." But just how was that agenda to be advanced?

The CLT recommendations were largely built around two areas: carbon pricing and emissions reductions, the warp and weft of the canvas. Neither can stand alone: warp without weft is just a dysfunctional pile of threads. The team specified that BC's carbon tax should be increased annually at $10 per tonne of CO_2 emitted, beginning in 2018, at the end of the five-year freeze that Clark had imposed in 2013. But in keeping revenue neutrality firmly in the foreground, the team recommended that a major share of the revenue from the carbon tax increase be married to a one-point cut in the provincial sales tax (PST), from 7 to 6 per cent, thereby continuing "BC's strong support for families and businesses through revenue recycling."

Gordon Campbell would almost certainly have approved. But Clark's government was hesitant, saying it would only consider an increase in the carbon tax "where emission-intensive, trade-exposed industries are fully protected from any carbon tax increase. In January 2016, British Columbia will begin consultations with industry and the public to consider new fiscal policies that would maintain the principle of revenue-neutrality ... fully protect our EITE sectors and maintain the competitiveness of BC businesses." Translation: we're unwilling to commit until the business community has had the chance to fully weigh in, even though that community has been well and ably represented on the Climate Leadership Team.[34]

The rest of the Clark government's press release was similarly milquetoast. In a key recommendation, the CLT advocated sectoral greenhouse gas reduction goals fifteen years hence, of 30 per cent for transportation and industry and 50 per cent for the built environment. All were (and still are) entirely achievable targets—if progressive carbon pricing is put in place with incremental investments in energy and transportation technology, retrofitting and renovation. Moreover, the CLT report recommended a suite of actions that would have the natural gas/LNG sector take fullest advantage of opportunities to reduce emissions by applying leading-edge technologies that would see up- and downstream methane leakage largely eliminated, and electric-drive compressors in common use. BC Hydro would be instructed to develop a generation and transmission strategy that would provide clean electricity in support of LNG production.

None of these recommendations were made frivolously; all were discussed at length and shaped with the benefit of extensive economic modelling and assumed application of leading technologies. The team members were confident that the combination of carbon pricing and emission-reductions directives would reinforce BC's reputation as a climate-action leader. That was something that Christy Clark had specifically asked them to do.

When asked almost six years later about the CLT recommendations, Clark praised the report as being definitive. "I thought it was a road map for us," she said. "I thought it provided the elusive balance that we've always been striving for between recognizing the need to fight pollution, fight carbon emissions, but not impair the economy in doing so."

But back in November 2015, the government wobbled in its willingness to adopt that road map. It refused in its press release to hammer a stake into the ground, saying only: "The province commits to a detailed review and analysis of the CLT's recommendations as well as other actions to develop a robust plan to establish and achieve emissions reductions across the three broad sectors identified by the team." For those of us on the Climate Leadership Team, that was a skinny bone on which to chew.

It didn't take long for that bone to fall to the floor. The government initially committed to having a draft climate-action plan—a response to the CLT recommendations—by early December, in advance of the Paris climate talks that month. A final plan was to be compiled and released by the Clark administration by March 2016.

But as a number of us on the leadership team noted in a letter to the premier on May 16, 2016, the "draft plan was cancelled and the deadline for the final plan was pushed to June."[35]

We were wrong. It was worse. The release date got pushed to late summer.

Clark and Environment Minister Mary Polak finally announced, on August 19, 2016, the release of the Climate Leadership Plan (CLP). The venue was a construction site, the future home of the Carbon Capture and Conversion Institute on Mitchell Island, an industrial enclave in the Fraser River on the southernmost margin of Vancouver, relatively far from reporters' offices and television studios. Any Friday afternoon release by any government is suspect, a signal that burial is preferable to publicity, but this was more than a routine interment. It was near the end of the dog days of summer, at the peak of the holiday season—and, as it turned out, the hottest day of 2016 in the Vancouver region. Clark's team had promised a "robust plan." What we got was anything but: a limp and submissive plan that whittled around the edges of climate action without striking at its heart.

The Climate Leadership Plan went nowhere near key—and bold—CLT recommendations, like resuming carbon pricing and lowering the provincial sales tax. Instead, it focused on the small scale: incentives to purchase electric vehicles; the construction of additional rapid transit lines; the conversion of ship engines to run on natural gas; the enhancement of efficiency requirements for gas fireplaces and air-source heat pumps. Twenty-one actions were announced, addressing "eighteen of the Climate Leadership Team's thirty-two recommendations," the press release crowed.[36] But what it didn't say was that among the fourteen recommendations *not* addressed were the keystones. There was to be *no* movement on carbon pricing in the new plan.

Instead, the government defaulted to wait-and-watch, admitting that while "any effective price signal on carbon has to go up over time," that would happen only "once others catch up." The premier and her team waffled, even as the world was so rapidly warming around them. "She just stonewalled," remarked esteemed Simon Fraser University economist Nancy Olewiler, a member of the CLT. Another member, environmentalist Tzeporah Berman, described the plan as "pathetic." The "number of our thirty-two recommendations accepted in full: zero," she wrote on Facebook.[37]

Criticism elsewhere could be particularly severe. Graham Thomson, writing in the *Edmonton Journal*, alleged that British Columbia had ceded climate-action leadership.[38] The province was "standing still," he wrote. The BC director of the Pembina Institute, Josha MacNab, zeroed in on the potential impact of the CLP: "The commitments in the plan represent a piecemeal approach that the Climate Leadership Team warned would prove economically and environmentally ineffective."[39] Simon Fraser University energy economist Mark Jaccard, a former climate adviser to Gordon Campbell, was particularly blunt: Premier Clark "has not taken any action on climate," he said.[40] "While she was talking about BC being a leader and her being a leader by inference, she was actually the antithesis of a leader, so that puts her in the cynical category."

So, what happened? Why the obviously weak response? That question touches a particularly raw nerve with Mary Polak. She freely admits the government was politically stuck when it came to increasing the then-frozen carbon tax. But accepting the CLT's recommendation for a $10 per tonne annual bump in the tax just wasn't going to fly, she insists. She would not reveal names, but says she had candid conversations with environmentalists, in which she asked, "No matter all this other stuff we're doing, we could do none of it and just raise the carbon tax as much as you guys want, and then that'd be okay?" She was told yes, it would. Polak thought that was foolish, retorting: "Do you realize how hard it is just to get this thing moving again? Like, don't give me a target, I can't reach. Just unfreezing the carbon tax is a big deal, right?" As she sees it, the intransigence of those unidentified environmentalists "broke my heart, because they were willing to have a political fight instead of supporting us." She wasn't happy when it came time to release the CLP.

"Were you just trying to keep your head down?" I ask. "Yes, because they weren't going to accept anything except total victory, which meant a huge jump to make up for how long it had been frozen. And I'm, like, you can't do that to people. What are you going to win? Because in the end, it'll just ensure we don't get elected again."

But the Climate Leadership Team wasn't proposing "a huge jump." The need to restore the upward ramp in the carbon tax was clear. It had proven successful in the five years following 2008. It was to be incremental, with relatively small annual increases, compensating fiscal supports and revenue neutrality. It was the right thing to do, and all it required was a willingness to reassume climate leadership, to get back to leading the parade, not following it. The evident unwillingness to act that Friday afternoon in August cannot be laid at the feet of committed environmentalists.

Another concern was plaguing the government that summer: the absence of a cap-and-trade program in jurisdictions outside British Columbia. That reintroduced the competitive-disadvantage argument—which was a serious constraint, remembers Polak: "One of the things [people] forget is when [the carbon tax] was originally designed, it was meant to work in concert with a cap-and-trade policy that everybody expected was coming, like across North America." By mid-2016, that policy had yet to appear at scale. "Our tax wasn't designed to exist on its own," she says. As a result, "we were going to have to make some changes to how the resource sector in particular, and other industries, were dealt with as far as the carbon tax." That position implies the carbon tax was having a negative impact, but in fact, given legislated revenue neutrality, it represents a distortion of the true economic bottom line.

Nevertheless, the political and economic concerns of the resource sector were an issue in 2016, and the Clark government was keen to finesse them. The nearly singular focus on natural gas and the fiscal windfall that it promised British Columbia led to significant speculation that the crafting of the Climate Leadership Plan was unduly influenced by the fossil fuel industry. Increasing and expanding the coverage of the carbon tax—to include, for example, fugitive methane emissions, as recommended by the CLT—would run the risk of "deterring the LNG industry," suggested Green Party Leader Andrew Weaver. "It's incompatible to be a climate leader and an LNG leader," he said.[41]

Environment Minister Polak didn't hesitate to push back. She defended the decision not to raise the tax, saying then that BC is "firmly in the lead when it comes to carbon pricing."[42] Pointing the finger at her critics, she said, "Even Alberta's plan—which many of these same individuals were praising not that long ago—doesn't get them to the level of pricing until about 2020." Polak was particularly concerned about potential economic fallout from a carbon tax increase. "There really is no accurate economic modelling out there to tell us what that does in an environment where no one else is doing it."

That comment would have surprised the economic modellers who worked hand-in-hand with the Climate Leadership Team, because the carbon tax increases they recommended were *obliged to be offset* by decreases in the provincial sales tax. Revenue neutrality—tax shifting—was to prevail. The political urge to focus on just one side of a two-sided economic equation was on full display.

Polak is also keen to dispel the inference that the Climate Leadership Plan was in some way bowing down at the altar of LNG. "None of [that commitment] is making us reluctant. Remember, our LNG industry is going to be the only LNG industry in the world that is going to be paying $30 per tonne carbon tax," she says. That was true, of course, but it's again only half of the equation. If the CLT recommendations had been put into effect, that same industry would have continued paying reduced rates of corporate income tax as well as less provincial sales tax. The CLT plan was not an attack on the natural gas industry. It was, however, *an attack on increases on emissions.*

Another twist came a year later. Based on their analysis of documents obtained via Freedom of Information (FOIA) requests, Shannon Daub and Zoe Yunker wrote in the *Narwhal* in 2017 that "while the Paris talks were underway [in December 2015], the BC government launched a closed-door, three-month-long process to work jointly with the oil and gas industry to revise and rewrite the climate leadership team recommendations." There were reportedly "five rounds of meetings over three months with all the key corporate players, from oil and gas producers to distributors." Senior officials from the BC Ministry of Natural Gas Development, the Climate Action Secretariat and BC Hydro were present. The only member of the Climate Leadership Team present was David Keane, president of the BC LNG Alliance. According to the FOIA materials, the attendees typically met in Calgary at the offices of the Canadian Association of Petroleum Producers, and were organized into working groups: on the carbon tax; on methane emissions from the natural gas industry; and on prospects for electrification of the LNG industry.

PowerPoint presentations delivered in Calgary, and obtained via FOIA, defined the timeline in early 2016 for "consultations" with the oil and gas industry. Working Group 1, focused on the carbon tax, was given two key tasks: to model impacts of the recommended tax increase and its trajectory and "start language," and then to "finalize language." UBC professor and political scientist Max Cameron, cited by the *Narwhal*, suggested this approach smacked of "institutional corruption."[43] But in the same *Narwhal* piece, Laurie Adkin, professor of political science at the University of Alberta, offered a somewhat more nuanced perspective, suggesting that government consultation with industry is the norm. "I do not believe that any climate change plan has been written, to date, in which the major fossil fuel corporations have not 'directly crafted' the plan," she said.

When asked specifically about the notion of institutional corruption, Christy Clark says: "I don't remember CAPP having anything to do with this. That does sound like a headline from the Centre for Policy Alternatives." She says the ministry of natural gas development would regularly meet with Shell—a major investor in BC, and a key member of CAPP—and with other investors in British Columbia based in Calgary. "We should be meeting with employers and talking about the impact of government policy on them," she says.

Mary Polak doesn't mince words when asked the same question: "It's a bunch of nonsense. And I know, because we held the pen on most of it." When it is pointed out to her that the only member of the Climate Leadership Team present at the Calgary meetings was the representative of the LNG Alliance, Polak replies: "Well, who would you expect to be there?"—implying that nothing untoward went on. When asked how much influence the natural gas industry had on shaping the plan, she says: "No more than anybody else. I mean, once we got the [CLT report and recommendations], the rest of it was all discussions amongst cabinet ministers and how you could balance different interests." But anyone keeping score would look askance at an 18:1 participation ratio at multiple Calgary sessions in the first three months of 2016.

One point is very clear: holding frequent meetings between senior Clark administration bureaucrats and fossil fuel execs was pretty much standard practice. Indeed, the Canadian Centre for Policy Alternatives reported in 2017 that the top ten fossil fuel firms active in the province lobbied the BC government more than 19,500 times between April 2010 and October 2016, some fifteen times as frequently as environmental organizations.[44] But such standard practice included another angle: it was not just the BC Liberal administration and its civil servants who were being lobbied; Opposition New Democrat MLAs were also heavily targeted, according to the CCPA. The natural gas industry was practicing equal-opportunity lobbying, and it had sway—significant sway.

Despite its intensive efforts to craft a new industry in British Columbia that would benefit from a highly favourable tax regime, it was not all clear sailing for the LNG Alliance. Future profitability could not be guaranteed, given the highly competitive international market in which it operated. That was seen to be particularly true if an accelerating carbon tax was to be introduced—revenue neutrality notwithstanding. Moreover, there was no guarantee that the future price would compensate for the very high capital investment that was demanded by LNG production in British Columbia.

Clear sailing requires calm waters. There, the industry held a very big ace: the premier and her team were so keen to launch this new economic engine that other concerns like carbon taxes and global warming could be shunted aside. Clark justifies the latter by focusing on replacing coal as a fuel: "China and India have to get coal out. And the replacement right now is LNG. I would argue that natural gas is an absolutely vital ingredient of fighting climate change for the next decade." As to launching the economic engine, her Climate Leadership Plan of August 2016 said it all. Industry got its way. Carbon tax increases, despite being carefully offset by the Climate Leadership

Team's fiscal road map, were ignored. Gordon Campbell, had he been present, would have shaken his head in dismay. Terry Lake was not happy with the outcome. "I remember having the feeling that it was sham exercise for politics rather than real change," he says. Down Under, Tony Abbott—the carbon tax wrecker—would have gloated.

Years later, Merran Smith rues the consequences of rejecting the CLT's recommendations. "Today we're paying the price from that delay in climate action. It's even harder and more expensive to reduce our emissions now. And it has to be done in a more extreme way, with deeper shifts. The early carbon tax and low-carbon fuel standard that Gordon Campbell put in place [in 2008] had sent a signal to industry that they had to reduce emissions. Those initiatives actually strengthened the economy and created a lot of new jobs and new sectors in cleantech. A vibrant, innovative tech sector emerged here. Climate action is not just something that you do once. You need to keep it going."

You need to keep it going. Amen.

EPILOGUE: *In the provincial election of May 9, 2017, some eight-and-a-half months after the release of the Climate Leadership Plan, Premier Clark lost her legislative majority, winning just forty-three of eighty-seven seats. She had campaigned heavily on what Maclean's magazine described as a "narrow, unimaginative focus on limited government and job creation" that "left no room for innovative policy, like the carbon tax her predecessor, Gordon Campbell, introduced, a North American first."*[45]

The appeal of limited government aside, another pair of words played a significant role in the election: Kinder Morgan. Four months before the election, Clark had approved Kinder Morgan's Trans Mountain Pipeline expansion project, claiming: "The project has met the five conditions."[46] *Her approval followed the feds' provisional approval for the expansion, granted* the previous November. Environmentalists were furious, one writing in the Vancouver Sun: "The cynical, albeit clever, five-conditions con allowed Clark to mask her Liberal government's support for Trans Mountain, via a combination of faux fence-sitting and feel-good 'standing up for British Columbians' rhetoric, at least until the federal Liberal government had taken the brunt of the criticism for approving the project."[47]

In the final tally, the New Democrats took forty-one seats and the Green Party three. On the face of it, the results reflected multiple advance polls that pegged the race as being neck-and-neck between the two major parties, with the Greens, led by Andrew Weaver, possibly winning enough seats to hold the balance of power—which they did, with attendant implications for progressive climate policy. Martyn Brown, Gordon Campbell's former chief of staff, asserted that "were it not for the Clark government's obsession with fossil fuel development and its support of projects like Kinder Morgan … Weaver's party would not have won the seats it did. At least one hundred thousand previous BC Liberal supporters voted Green this election. They were largely motivated by Weaver's leadership and by his party's unflagging commitment to fighting those and other projects, to fighting for renewable energy and for a green economy, and to reclaiming BC's global leadership on climate action."[48]

In the immediate wake of her narrow loss, Clark stated that she'd remain in office, seek the confidence of the legislature and attempt to govern with a minority. But on June 29, her Throne Speech was voted down by a quasi-coalition of the NDP and the Greens. When her request to dissolve the legislature and call a new election was rebuffed by the lieutenant-governor, she resigned. NDP Leader John Horgan was invited to form the government. Horgan was sworn in as premier on July 18, 2017.

Thus ended Christy Clark's six-year interregnum on climate action. On April 1, 2018, the progressive annual $5 per tonne rise in the carbon tax was reinstated by the NDP government with the Greens' support, along with enhanced direct fiscal support to low- and modest-income British Columbians.[49] While the annual increase was just half what the Climate

Leadership Team had advocated, it was at least a step forward, but with a short-sighted, politically vulnerable downside: the new revenue went into the Treasury, not toward continuing the critically important revenue-neutrality obligation that had been established in 2008.

If there is a key message arising from Christy Clark's final electoral campaign, it's a simple one. In 2017, British Columbian voters chose to support climate action. They chose to support carbon pricing. Christy Clark designed her platform around a different set of choices. Gordon Campbell says, "I would argue that Christy Clark lost in 2017 because of her policies on climate. If she had maintained the policies we had, and built on them, the Green Party would not have got the 17 per cent [of the popular vote] they did because, effectively, she abandoned those policies."

CHAPTER 9

How to Sell a Carbon Tax
Three Stories

The politics of carbon taxes are poisonous.
PAUL KRUGMAN, 2022

IT WAS MID-NOVEMBER, 2012. BARACK OBAMA, A DEMOCRAT, HAD just been re-elected president of the US a week earlier; hope was again in the air. I was sitting at breakfast in the Hotel Vancouver opposite Obama's ideological antithesis, Bob Inglis, an arch-conservative former Republican congressman of the Fourth Congressional District of South Carolina.

The Pacific Institute for Climate Solutions had helped to bring him to Vancouver and Victoria to meet with politicians, bureaucrats, editorial boards, academics and the public. Inglis was a tall, likeable, articulate and thoughtful man who sported an easy manner and an infectious, broad smile. Politically, he was the polar opposite of Obama, except for one thing: both he and the president were advocates for carbon pricing that carried with it some form of revenue neutrality.

But Inglis was no longer a political representative. He explained to me over scrambled eggs that after having served the citizens of southwestern South Carolina for twelve years—six congressional terms—he'd been "primaried" in the leadup to the 2010 midterm elections. Inglis had been badly defeated by another right-wing Republican because he'd had the audacity to stand up at an outdoor all-candidates meeting and tell the many attendees present that "man-made global warming is real, and we need to combat it by putting a levy on carbon emissions." That single statement, he told me, scuttled his chance of serving a seventh stint in Washington.

The Fourth District in South Carolina was then arguably the most conservative seat in a very conservative Southern state. Ideologically, Inglis had been a good fit. He was proud of his 93.5 per cent rating from the American Conservative Union; he was pro-life, a staunch Christian, supported the Second Amendment and had little time for unions. But on climate change, he followed the facts, and argued that conservatives should hold emitters accountable with a carbon tax. His advocacy had been very public; two-and-a-half years before his loss in the primaries, he'd co-authored an op-ed in the *New York Times* with Arthur Laffer, the famous conservative economist. Unknowingly channelling Gordon Campbell, Inglis and Laffer wrote: "A climate-change bill withered in Congress this summer because families don't need an enormous, and hidden, tax increase. If the bill's authors had instead proposed a simple carbon tax coupled with an equal, offsetting reduction in income taxes or payroll taxes, a dynamic new energy security policy could have taken root."[1]

A day after our breakfast meeting, Inglis addressed the BC Liberal caucus at an early-afternoon private meeting in the legislature. He'd been invited to speak by Minister of the Environment Terry Lake. Lake first hosted a small group of us at a lunch in the legislative dining room, where we were ushered to a quiet table off to the side. That was in itself an eye-opener. The minister explained that it was better for us to be sitting apart from the large cohort of his colleagues at the long Liberal table in the middle of the room because "a number of them aren't in favour of dealing with global warming." That observation was offered more than four years after BC's climate action plan—and carbon tax—had taken effect and was shown to be working.

An hour later, Bob Inglis stepped up to the lectern. I took a seat quietly at the very back. About forty Liberal MLAs—the great majority of the caucus—were present. Premier Christy Clark and Minister for Natural Gas Development Rich Coleman were notably absent.

Inglis began his twenty-minute talk by reciting his uber-conservative credentials. It was an impressive list, and it set a tight context for his subsequent comments, which made the point that true conservatives had an obligation to deal with climate change. The solution for his country, he said, was to put in place a revenue-neutral carbon tax, just like Gordon Campbell had done in BC.

It was powerful. Here was an avowed arch-conservative Republican advocating that the United States follow British Columbia's lead. Inglis was convinced that a carbon tax did not just have to *be* revenue-neutral; it had to be *seen* to be revenue-neutral. Taxpayers had to see a return on the tariffs they would be paying.

Terry Lake opened the floor to questions, and immediately John Les, a former cabinet minister whose riding was in BC's Bible belt, leaped to his feet. Les is a barrel-chested man with a booming voice. "I can't believe what I've just heard," he thundered. "Everybody knows that global warming is a socialist conspiracy." Minister Lake was clearly embarrassed; from my seat near the exit door, I could see him flush. But Les's boorish behaviour didn't faze Bob Inglis. He was quite used to dealing with such uninformed outbursts. He looked Les straight in the eye and calmly said, "Well, I don't care what you think about global warming. What have you got against having more money in your pocket if you burn less carbon?" It was a brilliant way to capture the importance of revenue neutrality. Les spluttered something unintelligible in reply and sat down.

Inglis one, climate deniers zero.

Inglis faced a few other questions that afternoon, all puerile, from backbencher members of Les's global-warming-is-a-hoax camp. They made zero impression.

Finally, an intelligent question did emerge. It came from Mary Polak, then minister of transportation and infrastructure. Polak was keen to hear Inglis's views on carbon pricing's impact on the transportation sector. His lengthy reply was an intelligent and all-too-rare exchange between two thoughtful politicians, free of grandstanding or axe-grinding. The majority of the members in the room took notice.

Inglis two, climate deniers zero.

Bob Inglis's visit that day—a cross-border exchange between conservatives—remains a salutary moment in serving notice to centre-right politicians that BC's climate leadership was being taken seriously outside provincial boundaries. In 2012, we still had much of which to be proud.

At a ceremony in Boston on May 3, 2015, Bob Inglis was named the 2015 recipient of the John F. Kennedy Profile in Courage Award for his contributions on the global warming file. The citation was read out by Kennedy's grandson, Jack Schlossberg:

> *Bob Inglis is a visionary and courageous leader who believes, as President Kennedy once said, that 'no problem of human destiny is beyond human beings.' In reversing his own position and breaking with his party to acknowledge the realities of a changing climate and its threat to human progress, he displayed the courage to keep an open mind and uphold his responsibilities as a leader and citizen at the expense of his own political career. His evolution in thought, brave stand and continued dedication to tackling the single biggest environmental and humanitarian crisis of our time inspires us all.*[2]

Hear, hear. We need many more like him.

CHAPTER 9

—

"Hey Dad, where is it?" asked my son in the early fall of 2017, holding a yellow and white Government of Canada cheque with his name on it.

"Where's what?" I replied.

"The BC carbon tax refund you told me I get."

"Let me see that, please," I said, holding out my hand.

My son was a university student, who, like all modest-income Canadians, qualified for quarterly payments from Ottawa to blunt the regressive edge of the national goods and services tax (GST). British Columbia piggybacks on the same vehicle to return carbon tax rebates to its low-income citizens and families, four times a year.

"I mean, where's the beef?" my vegan son asked, smiling quizzically, and pointing at the cheque. "There's no carbon tax refund mentioned here."

GST-refund recipients receive cheques that list just a *single* dollar value. But the deposit value of the cheque actually consists of *two* governmental contributions: *provincial* BC carbon-tax compensation and the *federal* GST refund, which for bureaucratic convenience are added together, yielding just one number. So where was the carbon-tax rebate mentioned, even tangentially?

It took a moment to spy it. A camouflage expert could hardly have hidden it better. And many cryptologists would have admired the attempt—inadvertent as it was—to disguise the meaning of the text, printed in small, black sans-serif font in the upper right corner: "GST/BCLICAT credit," right below the dominant black Times Roman "Canada." Just off to the side, near the edge of the cheque, in a poorly reproduced, grainy grey, sat a diminutive BC logo, with "British Columbia" at its side. And just below that, as befits a bilingual nation, sat some more sketchy, greyish text: "Crédit pour la TPS/TMCRFRCB," superimposed on a stylized Maple Leaf.

BCLICAT translates to "British Columbia Low Income Climate Action Tax." The French translation is, well, longer—and worse.

Who knew? Not most British Columbians. When "BCLICAT" was shown to an experienced banker in Victoria—who frequently handled such cheques—and she was asked what it meant, she said, "Maybe it's French for GST." She unwittingly highlighted a concern that Martyn Brown expressed about revenue neutrality and rebates in 2008 when he asked, "How much are you going to see it?"

That's it. Four times a year from 2008 onward, a significant fraction of British Columbia's population received a cheque with two indecipherable acronyms. In terms of a governmental communications opportunity, it was a travesty, a failure to bring society increasingly onside.

"The politics of carbon taxes are poisonous," wrote economist and Nobel laureate Paul Krugman in 2022.³ But that doesn't have to be true. Carbon taxes are, perhaps, poisonous *if not offset by other tax reductions*, but as British Columbia so clearly demonstrated in the Campbell years, if the public (and the business sector) are compensated through tax relief elsewhere and/or direct subventions, like my son received, society is accepting.

Communicating how carbon-tax offsets are delivered to individuals, families and even corporations is a never-ending challenge. It can be a challenge for government ministers too. In May 2016, six months after the Climate Leadership Team had submitted its report, I attended a cabinet briefing on proposed climate actions and suggested that the carbon tax, still frozen at the time, should be increased. Shirley Bond, MLA from the northern riding of Prince George–Valemount and at that time a minister in charge of labour issues, replied, "My constituents will never go for that." When I pointed out that their income taxes should go down correspondingly, assuming continuation of revenue neutrality, she said, "They don't see that and I can't sell that." I persisted. "Then that's a communication challenge that your government needs to take on," I said. It fell on deaf ears; it was clear that there was little appetite for selling the concept of tax shifting.

There is a take-home message here: *Carbon taxes need not be poisonous.* But there is also an obligation and, frankly, BC failed to measure up. Successful application of a revenue-neutral tax demands that compensating offsets and cash payments be well-publicized. Citizens need to be constantly reminded that they are getting their money back. BCLICAT or TMCRFRCB stamped in small font on a government cheque doesn't cut it, in either official language. It's bad marketing, full stop.

Even worse, it sells the people short. Transparent, publicized revenue neutrality—think of it as cash back—is a reward, an incentive for citizens to pitch in to help solve the climate crisis. It makes them part of the solution. When we miss the boat on that, we miss building a better future.

—

Marketing indeed matters. In 2014, I was invited to discuss BC's carbon tax in Bogotá at the Latin American and Caribbean (LAC) Carbon Forum, sponsored by the World Bank. Every country from Mexico to the tip of Tierra del Fuego was represented in the cavernous conference hall. On day one, I described to the four hundred delegates how Gordon Campbell and his team had implemented a carbon tax, what its impact had been, and how it had been received. It caused a stir. Interest was high. The head Mexican delegate at the meeting, Juan Carlos Belausteguigoitia, said afterward, "We are evaluating the performance of our tax and promoting changes that will make it more efficient and effective, and the experience of BC will be very valuable.

Do not be surprised if we call you in the near future." It was good to hear. BC was seen as an exemplar.

One of the sessions on day two was a panel discussion on design and implementation of carbon taxes. As a panellist, I was sitting between the former environment minister for Costa Rica and the current Colombian minister of the environment. As the discussion wrapped up, the Colombian minister tapped me on the shoulder and said, "I have a problem. We want to introduce a carbon tax in Colombia, but we can't do it politically. We have terrible air quality in Bogotá. Our kids are getting asthma at a worrying rate. They are suffering. We must reduce emissions. But the phrase *carbon tax* is toxic. The public won't accept it. What can we do?"

I pondered for a moment. Borrowing from Bob Inglis's playbook, I offered a path forward. "When you walk into meetings with your constituents or the media, tell them, 'I have a proposal for you—I want to reduce the income tax you pay.' And then don't say another word."

He looked perplexed.

I went on. "Wait until someone, a reporter perhaps, asks, 'But you need revenue. From where will the pesos come?' And it's only then that you say: 'As you all know, we have a very serious air-quality problem in Bogotá. Our kids are suffering. We need to get rid of the pollution that's causing the asthma epidemic. Let's tax the polluters and give the revenue right back to you in the form of lower income taxes. If you pollute less, you'll have new money in your pocket.'"

I was simply recommending a Coles Notes version of what we did in British Columbia in 2008, albeit with one key difference. Rather than lead with "We're going to raise your taxes to combat pollution," which was essentially the rationale we gave in BC, I suggested to him a variant. Change the lede. Open with a positive, an antidote to poison: "We're going to lower your income tax. Let me say that again. We're going to lower your income tax." And then, and only then, tack on the fiscal key: "We'll do that by putting a tax on the pollution that we are all so desperate to be rid of. And by the way, did I say that we're planning to lower your income tax?" I ended my admittedly overly simple advice by emphasizing that at every possible opportunity, he needed to remind those in the room that they would see lower income taxes. "Say *that* over and over again," I offered. "It's a way to invite them to be part of the solution, and it offers a bonus: The phrase *carbon tax* need not be used."

"Ah," he said, "I see. We have not tried selling a tax like that; I'll consider it."

In 2016 Colombia introduced a carbon tax, one of the first examples of carbon emissions pricing in South America.[4]

It needs to be said again: clear, sharp, consistent and honest marketing matters. Governments: take note.

CHAPTER 10

And Now to Canada: Have We Not Yet Learned?

Axe the tax!
> Frequently repeated by CAROLE JAMES, left-wing leader of the Opposition, British Columbia, throughout 2008 and into mid-2009

Axe the tax!
> Frequently repeated by TONY ABBOTT, right-wing prime minister of Australia, from 2011 through 2014

Axe the tax!
> Incessantly repeated by PIERRE POILIEVRE, right-wing leader of the Opposition, Canada, throughout 2023 and ongoing in 2024

Scrap the crap!
> Said by BILL BLAIR, centrist Liberal cabinet minister, Government of Canada, in Question Period, March 18, 2024

CHAPTER 10

MORE THAN A DECADE AFTER THE POLITICAL DRAMAS RECOUNTED in this book, Mother Nature is yelling "Fire"—no, "FIRE!"— even as global political systems continue to spawn wannabe Neros who would prefer to fiddle: 2023 was by far the warmest year in human history, adding yet another data point to an annual parade of such records. Climate extremes now routinely shock in their intensity, with a direct monetary cost that borders on the unfathomable: Over US$3 *trillion* in damages to infrastructure, property, agriculture and human health have already slammed the world economy this century owing to extreme weather. That stunning number will likely pale in comparison to what is coming. The World Economic Forum, hardly a hotbed of environmental activists, now reports that global damage from climate change will probably cost some $1.7 trillion to $3.1 trillion *per year* by 2050,[1] with the lion's share of the damage borne by the poorest countries in the world.

And yet we fiddle.

In today's Canada, a decade or more after Tony Abbott misled Australia, there is more deception, this time national in scope, coming directly from the right-wing Opposition benches in Ottawa. In 2023, the populist Conservative Leader Pierre Poilievre adopted "axe the tax" as his new mantra. Ripping a page from Abbott's playbook, he has vowed to shape the next federal election campaign around that hackneyed rhyme, which remains as specious in 2024 as it was in 2008.

In the current Canadian context, as in Australia years ago, the phrase "axe the tax" offers both good and bad news. The good is that Canada now actually has a national carbon tax; the bad is that it is politically vulnerable to being axed. And to a large degree, the responsibility for both circumstances—good *and* bad—falls to the well-meaning Liberal federal government.

Soon after ousting Stephen Harper's Conservatives from power on October 3, 2015, newly elected Liberal Prime Minister Justin

Trudeau announced that his government would establish a national carbon-pricing system. He was true to his word: nearly a year later, his "pan-Canadian approach to pricing carbon pollution" was announced by Catherine McKenna, then minister of environment and climate change.[2] The pending legislation would ensure that all provinces and territories would have carbon pricing in place by 2018. On June 21 of that year, the comprehensive Greenhouse Gas Pollution Pricing Act was proclaimed—almost exactly ten years to the day after British Columbia's carbon tax had begun its successful run.[3]

The new national carbon pricing system had two parts: a direct tax on fossil fuel consumption—the "federal fuel charge"—and a blueprint for regulatory trading, the federal Output-Based Pricing System (OBPS), which was designed to apply primarily to large industrial emitters. BC's fingerprints were all over the federal fuel charge. Applying to those jurisdictions that did not have pricing systems already in place (unlike BC and Quebec), the national scheme began in 2019 at a modest rate ($20 per tonne of carbon dioxide emitted) scheduled to ramp up $10 annually to $50 per tonne in 2022. Later regulations set an eventual target of $170 per tonne in 2030, equivalent to about forty cents per litre of gasoline. Scheduling that benchmark price to start low and rise over many years was purposely designed to give both industrial emitters and individuals time to adjust to a lower-carbon-pollution future, just as Gordon Campbell had stipulated for BC in 2008.

The Output-Based Pricing System zeroes in on large emitters by granting industries credits to pollute, based on an industry-specific, performance-related benchmark. Specific credit limits are fixed, and every tonne of carbon-dioxide equivalent emitted beyond the limit is taxed, thus establishing a compelling incentive to emit less. With a nod toward revenue neutrality, federal tax revenues under the OBPS are returned to the jurisdictions of origin, earmarked to support additional emissions reductions as well as the development of cleaner technologies and processes.

Moreover, the design of the OBPS facilitates straightforward alignment with long-standing provincial initiatives like Alberta's Technology Innovation and Emissions Reduction System, which is rather churlishly touted by the current Conservative government of Alberta as having been designed to allow the province "to reduce emissions without interference from Ottawa."[4]

There is one other benefit to the federal OBPS structure: although emissions credit limits are intended to be marginally reduced annually to stimulate ongoing conservation, the OBPS also allows pollution credit limits to be adjusted *upward* should production or international trading circumstances change, a direct benefit to emissions-intensive, trade-exposed industries. Overall, the OBPS architecture spurs innovation while maintaining competitiveness and protecting against "carbon leakage" (i.e., the risk of industrial facilities moving from one region to another to avoid paying a price on carbon pollution).[5]

It is an accommodating policy that the Canadian Climate Institute suggests is working. Between 2015 and 2030, annual emissions from large-scale industrial sectors in Canada, like oil and gas, are projected to fall by about a third, compared with a scenario in which no new climate policies are introduced after 2015.[6] The federal fuel charge is projected to cut national emissions by an additional 8 or 9 per cent over the same time frame.

But in the fractious world of climate politics, smart policies will only succeed if they are accompanied by political acceptability. There is one feature of the federal legislation that rises above all others in fostering that success: the requirement that almost all tax revenues be returned to taxpayers or the home jurisdictions of industrial emitters. That stipulation sat at the very core of the carbon tax established by the Campbell government in BC a decade earlier, and it became a centrepiece, indeed, *the* centrepiece, of Trudeau's scheme. In the context of the federal fuel charge, tax revenues collected are reimbursed directly four times a year in eight provinces and two territories (BC, Quebec, and the Northwest Territories run their own—but federally aligned—schemes). That provision of revenue neutrality is sacrosanct.[7] At present in Canada, about 90 per cent of the fuel-charge taxes collected is rebated right back into the pockets of taxpayers.[8]

The impact of the rebates is significant. In 2024, a rural family of four in Pierre Poilievre's home province of Alberta, for example, will receive back $2,160, some $700 more than the average family will pay in direct carbon taxes.[9] That average family will receive the same rebate regardless of its energy consumption: if it emits less, it will pay less carbon tax directly and will therefore be even better off. The redistribution fits hand-in-glove with the simple approach Gordon Campbell visualized almost twenty years ago, which was conservative and conservation-oriented: "You choose. You save."

It's also fair. As the CBC's data journalist Robson Fletcher pointed out: "A family of four living in a four-thousand-square-foot home with three cars in the garage and an RV in the driveway gets the same rebate as a family of four in the same city living in a five-hundred-square-foot apartment and relying on public transit."[10] Thus, low- or modest-income families disproportionately benefit from the rebate scheme, and most receive back significantly more than they pay out. It is a brilliant design largely built on British's Columbia's 2008 template. It should have come wrapped in political popularity.

Only it didn't.

The Trudeau government must shoulder much of the responsibility for that failure. Since the inception of the Greenhouse Gas Pollution Pricing Act, the government has neglected to remind Canadians over and over again that they are getting their carbon taxes back. The government lost its way, focusing on the word *environment*

CHAPTER 10

rather than on the phrase that speaks more directly to voters: *revenue-neutral*. Rebates were the hook, the reward for contributing to a better environment by paying an additional tax.

Instead, as reported by Radio-Canada, for years direct deposits into householders' bank accounts were given various nondescript or indecipherable labels like "CANADA FED" or "DN CANADA FED" or "Climate Action Incentive Payments."[11] According to Shachi Kurl, president of the Angus Reid Institute, polling data collected in late 2023 from across the political spectrum show "a massive number of Canadians who either believe that they are not receiving a rebate when they are, or don't know if they're receiving a rebate or not." She described the Liberal government's promotion of one of its flagship policies as "a failure at the most basic level of retail political communication."

Steps have been taken to repair the damage wrought by poor carbon-price marketing: in February 2024, rebate payments—finally!—were labelled "Canada Carbon Rebate."[12] But such a basic communication imperative is arriving years late. That delay came with a big cost; it opened the door to malfeasant opportunists like Pierre Poilievre, giving him the leeway to stand, repeatedly, behind a large sign tacked to his podium: "Axe the Tax/Abolir La Taxe," it says. There is no footnote on that sign to point out that tax monies are rebated directly to Canadian households. Moreover, Poilievre *never* stands behind that podium and points out that most families get back more than they pay, and if they take steps to reduce their carbon emissions, they will *still* get the same quantum of cash directly deposited quarterly into their bank account.

On the flip side of the rebate coin, has a single Canadian *ever* heard Poilievre publicly announce that rebates will disappear if he axes the carbon tax, and that almost all modest-income Canadians will be worse off? Has any parent with two kids in rural Alberta *ever* been advised by the Calgary-born leader of the Opposition that he would pull some $700 or more of extra cash per year from their pockets to pay for his sonorous axe-the-tax folly?

No.

Worse, Poilievre avoids all challenges that query the gaping hole in his fiscal logic: how would he replace the nearly $12 billion per year that the Canada Revenue Agency currently receives and redistributes through its tax on greenhouse gas emissions? He refuses to answer, instead belittling seasoned press gallery reporters who have the audacity to ask such tough questions, peppering them with condescension. A *Toronto Star* editorial recently described his approach this way: Poilievre's "media strategy is to denigrate, disrespect and vilify journalists who ask him questions he doesn't like, then weaponizing social media to amplify those encounters."[13]

Poilievre has adroitly avoided being held to account. He sits in an information vacuum of his own creation that has left far too many Canadians misguidedly thinking that somehow they'll be in a better financial position if the carbon tax is axed. The opposite is true, but who can blame them for thinking otherwise? Aided in part by a sleepwalking Liberal government, Poilievre has been given nearly free rein to misrepresent Canada's carbon pricing system.

There are signs that the federal government has finally acknowledged its own deficiencies and pushed back on the outright untruths coming from the Conservatives. On March 17, 2024, Jonathan Wilkinson, Canada's minister of natural resources, responded bluntly: "We do need to do a better job of communicating the affordability dimensions of the price on pollution because it is something that actually makes life more affordable … The facts are clearly that people get more money back. Mr. Poilievre should stop lying to Canadians. He is telling lies on an ongoing basis, and that's just not something that a responsible leader should be doing."[14] A day later, in Question

Period, Bill Blair, minister of defence, doubled down. "Perhaps it's time to scrap the crap," he said, alluding to the numbingly repetitive "axe the tax" and "spike the hike" sloganeering streaming from Poilievre's ranks.[15]

Repeated often enough, lies have traction. It's a cynical strategy, but for Pierre Poilievre it appears to be working. His popularity has risen dramatically since mid-2023. Skilled communicator that he is, he has effectively sold a new narrative, preying on Canadians' worries about inflation and the high cost of living, verbally linking them to the upward ramp in the carbon price. But analyses of real evidence—hard data from British Columbia and a suite of European jurisdictions—now confirm that his narrative is a false construct. Revenue-neutral carbon taxation systems do *not* contribute significantly to inflation.[16] Tiff Macklem, governor of the Bank of Canada, made that clear in September 2023, stating that the $15 per tonne annual increase in the carbon tax would add just 0.15 per cent per year to the rate of inflation in Canada. It's a "relatively small effect," he said.[17]

Herein lies a challenge for both the federal Liberal government and the mainstream media: Don't let Poilievre's lies gain further traction. Expose them. Throw down a verbal spike belt. Puncture and deflate his treadless tires with prose that can be easily understood. This won't be easy, given Poilievre's avoidance of questions, but illuminating mendacity is critical. There is too much at stake to let lies fester.

Unfortunately, the federal Liberals have repeatedly shot the puck into their own net. Public perception of revenue neutrality, for example, was enfeebled in past years by camouflaging carbon rebate cheques with obscure acronyms. That was the first own goal. Then, in the late fall of 2023, the prime minister announced that his government would carve out a three-year-long federal carbon-tax exemption for home heating oil in the Atlantic provinces, a time span that conveniently extended at least a year beyond the likely date of the next federal election. That manoeuvre, designed to curry voter favour in the Maritimes, spectacularly backfired. It immediately put a crack in a foundational element of Canadian carbon pricing: there can be no carve-outs for specific emitters or specific regions, no free riders, a requirement that Gordon Campbell and Carole Taylor had made crystal clear in British Columbia in 2008. Pierre Poilievre pounced on Trudeau's vote-buying gaffe, writing: "After plummeting in the polls, a flailing, desperate Trudeau is now flipping and flopping on the carbon tax."[18]

Completing an unwelcome hat trick was the puck slapped almost immediately into the net by premiers from Ontario and the West, who demanded that their provinces, too, receive such largesse from the federal Liberals. That demand—for an exemption not just on heating oil but also on natural gas—yielded intense media coverage across the country, which in the closing months of 2023 capped off a damaging barrage of attacks on Canada's carbon tax policy. Sadly lost in the frenzy was the real victim of the Liberals' self-immolation: the core concept of revenue neutrality that continues to put about $1,000 annually into the pockets of every taxpaying family in the Maritimes, a subvention that will rise year after year to compensate for the accelerating carbon tax. *That* should have been the lede.

Trudeau was wrong not just in granting a carve-out but in tackling the wrong target. It is the world market, and not the carbon tax, that is driving up the cost of home heating for the many thousands of households in Atlantic Canada still warmed by burning oil. The retail price of a litre of furnace oil in Halifax at the end of December 2019 was $1.05.[19] Nearly three years later it was an extraordinary $2.27, and by December 2023 down to $1.63. After five years of existence, the carbon tax added only seventeen cents per litre to that price. No, the carbon tax did not cause the affordability problem—the up-and-down world oil market did. The prime minister and his team should thus have taken aim at the real challenge: ridding the Canadian landscape of oil furnaces. They are yesterday's technology and they burn yesterday's fuel. Meeting that transformation is the wheel to which the Liberals should have put their shoulder. Ramped-up provision of low-interest, long-amortization federal loans for the purchase and installation of heat pumps would have been a progressive, inexpensive and smart solution to help address space-heating affordability concerns. Singling out home heating fuel for a carbon tax pause was no solution at all.

And what is Pierre Poilievre's position on heat pumps versus oil furnaces? His stance on this issue is invisible. The Conservative Party of Canada's website, as of spring 2024, offers not a word, *not one word*, on the party's carbon pricing positions.[20] But an undated and outdated website from the previous leader, Erin O'Toole, does, stating on page three: "We recognize that the most efficient way to reduce our emissions is to use pricing mechanisms."[21] Good, and true, so far. And then a little later: "Carbon pricing should not result in Canadians sending billions of dollars of new tax revenue to the government—revenue which it will be increasingly tempted to spend." Bad, and purposely misleading. Nowhere to be found is the phrase *revenue neutral* or the word *rebate*. And therein lies the very nub of a growing political challenge in Canada: a large fraction of our current political landscape—a fraction that is almost entirely conservative—refuses to recognize, even in the face of overwhelming evidence, that revenue-neutral carbon pricing is intelligent policy, as the conservative-leaning editorial board of Canada's national newspaper, *The Globe and Mail*, argued in mid-March 2024.

A few days later, a cohort of conservative premiers from Alberta, Saskatchewan and New Brunswick met virtually with a federal parliamentary committee to push for a pause on the scheduled $15 per tonne carbon tax increase scheduled for April 1. The absurdity of the premiers' position was immediately evident. Saskatchewan Premier Scott Moe was first up.[22] He was particularly concerned with capital investment and productivity growth in his home province, stating at one point, without offering any evidence, that "the price on pollution is creating uncertainty in the investment environment in Saskatchewan and Canada." But carbon tax increases are fully scheduled to 2030. Businesses and individuals factor those rises into their forward-looking corporate and household budgetary plans ("You choose, you save"). In his zeal to pause the April 1 increase, Premier Moe apparently did not see that if he convinced the federal government to forgo the long-scheduled carbon tax increase, he would be a primary agent in creating the very uncertainty he was decrying.

Alberta's right-wing Premier Danielle Smith, in her opening remarks, launched into an attack on the carbon tax generally, the scheduled April 1 increase, and the upward ramp to 2030: "This isn't just reckless, it's immoral and it's inhumane," she said.[23] "The added pressure will ruin countless lives, futures and dreams. It's a weight that Canadians can't bear. And that's why Alberta has been calling on the federal government to eliminate the carbon tax since 2019." Missing from her hyperbolic rhetoric was the phrase *revenue neutrality*. That evening, Andrew Coyne, of CBC-TV's "At Issue" panel, zeroed in on that omission, describing Smith's remarks as a "parade of nonsense" and "completely dishonest."[24]

In questioning from committee members, Liberal MPs focused on Smith's duplicity. She was hoist on her own petard by Ontario MP Irek Kusmierczyk when he asked her how she justified railing against the pending three-cent jump in the federal fuel charge on gasoline, when her own provincial budget, set to land on the same day, April 1, was scheduled to increase the provincial tax on gasoline by *four* cents a litre.[25] With a face that would do justice to a Texas hold 'em champion, Kusmierczyk asked the premier if she could clarify whether or not the axe-the-tax rally she had attended the day before was "a rally to axe *your* tax which is adding four cents to a litre?" Smith responded with a word salad, mentioning buzzwords like "home heating oil," "inflation crisis" and "building roads," but failing to address her hypocrisy.

Chantal Hébert, on CBC's "At Issue," sarcastically wondered if Smith was suggesting tax increases are "inhumane and immoral" only when they are federal, while provincial increases are immune to such value judgments. When the premiers were asked for their alternatives, what they offered to the committee was "just fantasy," said an incredulous Coyne, adding: "I think we saw how bankrupt their positions were." According to Coyne, when Saskatchewan Premier Scott Moe was asked about ways to limit emissions other than carbon pricing, he replied, "We looked at all the alternatives and they all cost more than the carbon price." "Gosh," exclaimed Coyne on the CBC, "I wonder what we should conclude from that?"

If anyone ever seeks an example of thinking as twisted as a corkscrew, they need look no further than the conservative premiers that week. They simply refused to acknowledge the central importance of revenue neutrality as a key to successful carbon pricing. They refused to acknowledge that the Canadian system, started by British Columbia nearly twenty years ago, is fundamentally a smart tax shift that benefits the economy *and* helps to drive emissions downward. In refusing to recognize that simple fact, they promise to take us backward in confronting global warming. And *that* is truly immoral and inhumane.

Australia's emissions reeled from such a backward lurch. Julia Gillard's carbon tax on large emitters—the foundation of her Clean Energy Act—was in effect for only two years before Tony Abbott repealed it in July 2014, just eleven months after he won the September 2013 election. Carbon dioxide emissions fell in 2013 by some 2.4 per cent, and were on track for a repeat performance until Abbott's guillotine fell. Within months, the declining emissions profile spun around; CO_2 emissions rose dramatically in 2015 and 2016. Australia squandered an opportunity that would have had an environmental and political impact on the global stage. Charles Komanoff, writing in 2020 for the Carbon Tax Center in New York, astutely observed what might have been:

> *The example of a proudly macho and avowedly hedonistic society explicitly pricing its carbon emissions could have turned heads, particularly in the United States, with its many cultural resemblances to Down Under. The fantasy of a land that produced Mad Max and Crocodile Dundee serving as a global role model for carbon taxing and reduction is delicious, though alas, [now] no more than that.*[26]

Democracy is a capricious lady. When she elected Abbott, she thrust a spanner into Australia's climate-action policy. There is an electoral corollary here that applies to Canada: modern elections in well-established democracies unavoidably have climate-action consequences. Full stop.

There's a second corollary, at least as important: wise leadership matters—leadership guided by history, science, economic experience and example. It was almost two decades ago that British Columbia accepted the reality that human activities are changing the Earth's climate at a rate that threatens both natural and socio-economic systems. The province acted on that implicit challenge then, imposing an escalating price on carbon emissions, lowering taxes, setting tough emissions-reductions targets and low-carbon fuel standards, demanding that public institutions and municipal operations become carbon-neutral, and accelerating investments in greener buildings and electrified public transit, all the while promoting renewable energy and technology development. We led.

That was then. A severe international recession, and changes on the federal and provincial political landscapes, stole wind from BC's sails. Vision faltered. By 2013 the province reverted to its historical default: exploitation of natural resources as the primary economic driver. The internationally celebrated revenue-neutral carbon tax was frozen. As LNG became the new darling, rules were changed to allow the natural gas industry to increase carbon emissions as it compressed gas for export. Job creation in the renewable energy and cleantech sectors slipped down the priority list. We stalled.

Things were different in the three Cascadia states immediately to the south. Jay Inslee, then newly elected as governor of Washington State, said on January 16, 2013, "The world will not wait for us," and that the state faced no greater challenge "than leading the world's clean energy economy." In 2015, he signed into law the Carbon Pollution Accountability Act, which imposed a price on emissions from large sources while incentivizing clean-energy innovation. Inslee's remarks had echoed those of John Kitzhaber, governor of Oregon, who, in releasing his ten-year energy plan in 2012, said it "provides a framework to move Oregon beyond a boom/bust economic cycle that depletes our natural capital and leaves us vulnerable to fluctuations in global markets, and moves us toward a future where our state is a leader in energy efficiency, home-grown renewable energy resources, and clean energy employment." And in California, thanks to Assembly Bill 32, signed into law in 2006 by the Republican Arnold Schwarzenegger, Governor Jerry Brown was able to observe in his State of the State speech on January 24, 2013: "By 2020, we will get at least a third of our electricity from the sun and the wind and other renewable sources—and probably more." Much of that was expected to be supplied by homegrown technology.

More than ten years ago, those three governors understood that the world would not wait for them in seizing the opportunity presented by global warming. It would not wait for BC either, but in the Christy Clark years the decision was made to turn away from vigorous and courageous promotion of a low-carbon future, just as in Australia. The province shied away from what Gordon Campbell had described in 2008 as "the long-commitment challenge of our generation," even as, relative to the rest of Canada, emissions had been declining while the economy was continuing to prosper.

We must never forget that central to meeting "the long-commitment challenge" is the need to put a price on carbon. "There is simply no other policy that is as important," wrote UBC economics professor Werner Antweiler in 2021.[27] Where properly designed, with revenue neutrality and progressive acceleration as its core elements, it works. We showed that in British Columbia well over a decade ago, and we did so with public acceptance—grudgingly, perhaps, but present nonetheless—despite our poor, almost invisible efforts to remind citizens that they were getting the tax back. The intelligent, hard-working vintner in the Okanagan Valley who complained about the cost of putting gas into his pickup, blaming "that goddamned carbon tax," was owed reminders that farm gas was exempt and that his personal income tax had declined, as had the corporate income tax paid by his thriving winery. While in Opposition, left-wing political leaders who were bleating about the tax being an imposition on the poor needed to be reminded that in fact BC's carbon tax was revenue-neutral, and that lower-income British Columbians and families typically received more in direct cash carbon-tax rebates and supports than they paid in consuming fossil fuels.

Now, more than ever, Canadians need to be reminded that the same is true nationally. Families in Canada sacrifice by paying a carbon tax, but they are rewarded by getting much or more of their outlay back *while directly contributing to climatic responsibility*. That contribution is an outcome worth celebrating.

Yes, BC did it right some sixteen years ago. And Justin Trudeau got it largely right on the federal stage in 2018. While both jurisdictions fumbled on the marketing front, we showed the larger world how to bring a carbon tax into play. Ours works—it *is* contributing to emissions reductions. The Abbott administration in Australia failed to learn from the Canadian experience, despite having British Columbia's data in hand, a fact that many Australian progressives still rue today. Unvarnished political opportunism that sold a carbon tax as poison triumphed in that nation. Intelligent, reasoned discourse was steamrolled by bellicose, belligerent whitewashing and the powerful Australian coal lobby. Reason failed to cut through negative rhetoric.

As these final paragraphs are being written, Australia's dismal carbon pricing history provides a lesson for Canada. We are on the verge of seeing that history be repeated here, a combination of tragedy and farce.[28] The parallels are stark. Tony Abbott and Pierre Poilievre are conservative birds of the same feather: one relished—and the other now relishes—the prospect of wielding an axe. Neither voice, ten years ago or today, offered—or is offering—a rational, coherent, economically viable solution that will help meet the global warming challenge.

But someone else did. A single province on the westernmost side of Canada stepped up and showed all nations that fair, redistributive, broad-spectrum carbon pricing could be done and done well, without economic harm. British Columbia did it.

It took wisdom. It took commitment. It took vision. It was transformational.

And it remains a template for the world.

Let us not turn our backs on that success now.

Endnotes

Prologue

1. T. Bolch, B. Menounos, and R. Wheate, "Landsat-based inventory of glaciers in Western Canada, 1985–2005," *Remote Sensing of Environment* 114 (2010): 127–37.
2. J. Tyndall, "On the absorption and radiation of heat by gases and vapours," *Philos. Mag.* 22 (1861): 169–94 and 273–85.
3. W.S. Broecker, *Science* 189 (1975): 460–63, DOI: 10.1126/science.189.4201.460.

Chapter 1

1. L. Safranyik and B. Wilson, eds., *The Mountain Pine Beetle: A Synthesis of Biology, Management, and Impacts on Lodgepole Pine* (Canadian Forest Service, 2006).
2. L. Gawalko, "Mountain Pine Beetle Management in British Columbia Parks and Protected Areas," in *Mountain Pine Beetle Symposium: Challenges and Solutions*, eds. T.L. Shore et al. (Canadian Forest Service Information Report BC-X-399, 2006), 298 p.
3. David Suzuki used this descriptive phrase to describe the pine beetle onslaught in the CBC-TV *Nature of Things* episode "The Beetles Are Coming," first aired on August 23, 2014.
4. Personal communication with the author, February 17, 2017.
5. Personal communication with the author, March 27, 2017.
6. L. Safranyik and A.L. Carroll, "The biology and epidemiology of the mountain pine beetle in lodgepole pine forests," in eds. L. Safranyik and B. Wilson, *The Mountain Pine Beetle: A Synthesis of Its Biology, Management and Impacts on Lodgepole Pine* (Natural Resources Canada, Canadian Forest Service, Pacific Forestry Centre, Victoria, BC, 2006), 3–66.
7. Safranyik and Linton (1991), cited by Safranyik and Carroll, "Biology and epidemiology."
8. Pacific Climate Impacts Consortium. Data available at: https://pacificclimate.org/data/bc-station-data.
9. Cathryn Atkinson, "Artisans promoting silver lining in blue pine," *Globe and Mail*, January 17, 2006, http://www.theglobeandmail.com/news/national/artisans-promoting-silver-lining-in-blue-pine/article965072/.

10. Yukon Ministry of Energy, Mines and Resources, "2015 Yukon Forest Health Report" 40 pp., https://emrlibrary.gov.yk.ca/forestry/forest_health/2015.pdf.
11. Natural Resources Canada, "Mountain pine beetle (factsheet)," http://www.nrcan.gc.ca/forests/fire-insects-disturbances/top-insects/13397.
12. A.L. Safranyik et al., "Potential for range expansion of mountain pine beetle into the boreal forest of North America," *Canadian Entomologist* 142: 415–42.
13. Safranyik and Wilson, 94–95.
14. BC Ministry of Forests, Lands and Natural Resource Operations,"Major Primary Timber Processing Facilities in British Columbia 2014," https://www2.gov.bc.ca/assets/gov/farming-natural-resources-and-industry/forestry/fibre-mills/mill_report_2014.pdf.
15. *Kamloops This Week*, "Merritt mill closure latest in series of Interior setbacks going back 15 years," September 24, 2016, https://archive.kamloopsthisweek.com/2016/09/24/merritt-mill-closure-latest-in-series-of-interior-setbacks-going-back-15-years/.
16. BC Ministry of Forests, Lands and Natural Resource Operations, "A Competitiveness Agenda for BC's Forest Sector," 2016, https://www2.gov.bc.ca/assets/gov/farming-natural-resources-and-industry/forestry/competitive-forest-industry/print_version_bcfs_agenda_final_lrsingles_r2.pdf.
17. Stumpage fees are like royalties. The government collects a fee for each tree that is harvested in BC, where 96 per cent of forested land is owned by the province.
18. G.F. MacDougall, "Water Storage of Pine: A Strategy to Mitigate Losses Due to Pine Bark Beetle" (MBA thesis, Simon Fraser University, 2005), 76 pp.

Chapter 2

1. Interview with the author, July 27, 2017. Except where otherwise indicated, all quotations attributed to Mr. Campbell in this book stem from that interview.
2. Mike Harcourt and Ken Cameron, with Sean Rossiter, *City Making in Paradise: Nine Decisions That Saved Vancouver* (Douglas & McIntyre, 2007), 113.
3. Ibid., 115.
4. Ibid., 117.
5. *Creating Our Future: Steps to a More Livable Region* (technical report), "Choices Make a Difference" (Burnaby: GVRD, 1990). Available at Metro Vancouver Corporate Library, call no. GVRD01096.
6. Harcourt et al., *City Making*, 121.
7. Ibid., 127.
8. H.L. Ferguson, "Twentieth Anniversary of the Toronto Conference on Our Changing Atmosphere: Implications for Global Security," *CMOS Bulletin* 36 (2008): 159–61.

ENDNOTES

9   Task Force on Atmospheric Change, "Clouds of Change" (City of Vancouver, 1990), 93 pp., https://a100.gov.bc.ca/pub/eirs/finishDownloadDocument.do?subdocumentId=3851.

10 Interview with the author, June 16, 2017. Except where otherwise indicated, all quotations attributed to Mr. Brown in this book stem from that interview.

11 BC Liberal Party, "A New Era for British Columbia," 36 pp., https://www.poltext.org/sites/poltext.org/files/plateformesV2/Colombie-Britannique/BC_PL_2001_LIB_en.pdf.

12 James G. Speth, *Red Sky at Morning* (Yale University Press, 2004). A co-founder of the Natural Resources Defense Council and later chair of President Jimmy Carter's Council on Environmental Quality, Speth wrote the book while serving as dean of the Yale School of Forestry and Environmental Studies. He chose the sailor's warning in the title because he foresaw "big trouble ... coming down the pike—and coming fast indeed" (p. xi).

13 Ibid., xi-xii.

14 Ibid., 161 (italics added).

15 Michael M'Gonigle and Justine Starke, *Planet U: Sustaining the World, Reinventing the University* (New Society, 2009), 288 pp.

16 Legislature of California, Assembly Bill 32, 2006, 13 pp., http://www.leginfo.ca.gov/pub/05-06/bill/asm/ab_0001-0050/ab_32_bill_20060927_chaptered.pdf

17 Government of British Columbia, Speech From the Throne, February 13, 2007, https://www.leg.bc.ca/content/legacy/web/38th3rd/Throne_Speech_2007.pdf.

18 Personal communication, September 20, 2017.

19 The Western Climate Initiative was initially a partnership of several western American states and Canadian provinces, including BC, that had a collective goal of reducing greenhouse gas emissions. A description is available at: https://archive.news.gov.bc.ca/releases/news_releases_2005-2009/2007OTP0053-000509.htm.

20 Government of British Columbia, Speech From the Throne, February 12, 2008, https://www.leg.bc.ca/content/legacy/web/38th4th/Throne_Speech_2008.pdf.

21 That plan was being actively drafted in the background by a trio of senior civil servants (see text) with input from an internal government body, the Climate Action Secretariat (CAS) and an external advisory group of experts, the Climate Action Team. CAS was created in mid-late 2007. Reporting directly to the premier, CAS comprised a group of talented senior provincial bureaucrats charged with helping to shape BC's climate action agenda. The Climate Action Team's members were announced on November 20, 2007. They were drawn from the university community, government research labs and the private sector, and Gordon Campbell described them as "some of the province's best minds," recruited "to help the government to aggressively reduce British Columbia's

greenhouse gas emissions by 33 per cent by 2020." (Source: http://vancouversun.com/news/staff-blogs/b-c-climate-action-team.)

22 Speech From the Throne, 2008, http://www.bcbudget.gov.bc.ca/2008/speech/2008_Budget_Speech.pdf.

23 Interview with the author, October 23, 2017. Except where otherwise indicated, all quotations attributed to Ms. Taylor in this book stem from that interview.

24 Janice MacKinnon, "The Green Shift, the Liberals and the west," *Policy Options*, November 1, 2008, https://policyoptions.irpp.org/magazines/les-elections-federales/the-green-shift-the-liberals-and-the-west.

Chapter 3

1 Paul Ekins at al., 2009, "The case for green fiscal reform: Final report of the UK Green Fiscal Commission," Green Fiscal Commission, London, 2009, https://westminsterresearch.westminster.ac.uk/item/90wyv/the-case-for-green-fiscal-reform-final-report-of-the-uk-green-fiscal-commission.

2 For a more technical discussion, see Thomas F. Pedersen and Stewart Elgie, *Critical Issues in Environmental Taxation* 15, eds. L.A. Kreiser et al., (Northampton, MA: Edward Elgar), 3–15.

3 Andy Robinson and Glen Armstrong, interview with the author, December 13, 2022.

4 *CBC News*, "New carbon tax receives praise, sparks criticism," February 19, 2008, https://www.cbc.ca/news/canada/british-columbia/new-carbon-tax-receives-praise-sparks-criticism-1.717093.

5 Christa Marshall, "British Columbia Survives 3 Years and $848 Million Worth of Carbon Taxes," *New York Times*, March 22, 2011, http://www.nytimes.com/cwire/2011/03/22/22climatewire-british-columbia-survives-3-years-and-848-mi-40489.html?pagewanted=all.

6 Canadian Press, "Carbon tax bumps up B.C. fuels prices," *CBC News*, July 1, 2011, https://www.cbc.ca/news/canada/british-columbia/carbon-tax-bumps-up-b-c-fuels-prices-1.1009207.

7 Eduardo Porter, "Does a Carbon Tax Work? Ask British Columbia," *New York Times*, March 1, 2016, https://www.nytimes.com/2016/03/02/business/does-a-carbon-tax-work-ask-british-columbia.html?_r=8.

8 Dirk Meissner, "B.C. businesses join call for carbon tax hike," *Global News*, March 30, 2016, https://globalnews.ca/news/2609301/b-c-businesses-join-call-for-carbon-tax-hike/.

9 British Columbia Statistics, "BC Gross Domestic Product at Basic Prices," BC Stats (2017), Chained 2017$ (1997–2023) and Current$ (2007–2020)

(May 2024). https://www2.gov.bc.ca/gov/content/data/statistics/economy/bc-economic-accounts-gdp.

10. Jock Finlayson, "Comment: Carbon tax has little impact on fuel consumption," *Times-Colonist*, August 24, 2013, http://www.timescolonist.com/opinion/op-ed/comment-carbon-tax-has-little-impact-on-fuel-consumption-1.599469.

11. Government of British Columbia, "Provincial inventory of greenhouse gas emissions," https://www2.gov.bc.ca/gov/content/environment/climate-change/data/provincial-inventory.

12. BC Stats, Government of British Columbia, "Forestry," data available at: https://www2.gov.bc.ca/gov/content/data/statistics/business-industry-trade/industry/forestry.

13. Jim Johnson, Pacific Analytics Inc., "The Impacts of the Carbon Tax on Vehicle Fuel Use in Metro Vancouver," 2015. Retrieved from https://web.archive.org/web/20160416134344/http://pacificanalytics.ca/node/46.

14. Nicholas Rivers and Brandon Schaufele, "Salience of carbon taxes in the gasoline market," *Journal of Environmental Economics and Management* 74, 23–36.

15. Andy Skuce, "The effect of cross-border shopping on BC fuel consumption estimates," *Critical Angle*, August 18, 2013, http://critical-angle.net/2013/08/18/the-effect-of-cross-border-shopping-on-bc-fuel-consumption-estimates/; and Yoram Bauman, "The Canadians are Coming! Cross-border fill-ups and the BC carbon tax," Sightline Institute, May 21, 2014, http://daily.sightline.org/2014/05/21/the-canadians-are-coming/.

16. Chad Lawley and Vincent Thivierge, "Refining the Evidence: British Columbia's Carbon Tax and Household Gasoline Consumption," *Energy Journal* 39 (2018), 147–71.

17. For a more technical discussion, see: Pedersen and Elgie, *Critical Issues*.

18. CBC News, "B.C. carbon tax jumps more than 1 cent," July 1, 2010, http://www.cbc.ca/news/canada/british-columbia/b-c-carbon-tax-jumps-more-than-1-cent-1.915792.

19. Canadian Press, "Carbon tax bumps up B.C. fuels prices," CBC News, July 1, 2011, https://www.cbc.ca/news/canada/british-columbia/carbon-tax-bumps-up-b-c-fuels-prices-1.1009204.

20. Shawn McCarthy, "B.C. to raise carbon tax, price of gasoline July 1," *Globe and Mail*, June 27, 2012, https://beta.theglobeandmail.com/report-on-business/industry-news/energy-and-resources/bc-to-raise-carbon-tax-price-of-gasoline-july-1/article4374532.

21. Ibid.

22. N. Rivers and B. Schaufele, 'The effect of British Columbia's carbon tax on agricultural trade," 2014, www.pics.uvic.ca/sites/default/files/uploads/publications/Carbon%20Tax%20on%20Agricultural%20Trade.pdf.

23 Statistics Canada, "Production and value of greenhouse fruits and vegetables," Table 32-10-0456-01 (formerly CANSIM 001-0006), release date April 25, 2024, https://www150.statcan.gc.ca/t1/tbl1/en/tv.action?pid=3210045601&pickMembers5B0%5D=1.12&pickMembers5B1%5D=4.1&cubeTimeFrame.startYear=2007&cubeTimeFrame.endYear=2013&referencePeriods=20070101%2C20130101.
24 Interview with the author, November 3, 2017.
25 Email communication with the author, November 3, 2017.
26 George Kondopulos and Lorne Burns, "British Columbia Cleantech: 2019 Status Report," KPMG, February 2020, https://assets.kpmg.com/content/dam/kpmg/ca/pdf/2020/02/bc-cleantech-2019-status-report.pdf.
27 Interview with the author, December 21, 2022. Unless otherwise indicated, all quotations attributed to Mr. Rhone in this book stem from that interview.
28 Email communication with the author, November 3, 2017.

Chapter 4

1 Kim Nursall, "War in the Woods mass arrests 20 years ago prompted lasting change," *CTV News*, August 11, 2013, http://bc.ctvnews.ca/war-in-the-woods-mass-arrests-20-years-ago-prompted-lasting-change-1.1406602.
2 Wallace S. Broecker, "Climatic change: Are we on the brink of a pronounced global warming?" *Science* 189 (1975): 460–63.
3 Anita Talberg et al., "Australian climate change policy to 2015: A chronology," Parliament of Australia, Research Paper Series, 2015–16.
4 Ross Garnaut, "The Garnaut Climate Change Review," 2008, https://webarchive.nla.gov.au/awa/20190509040128/http://www.garnautreview.org.au/index.htm.
5 Interview with the author, March 3, 2017. Unless otherwise indicated, all quotations from Dr. Garnaut in this book stem from that interview.
6 The Liberal Party in Australia is like what it was in British Columbia: liberal in name only, conservative in practice. In April 2023 the BC Liberal Party officially changed its name to BC United. Politically, it remains right of centre.
7 Robert Chynoweth's full speech can be found in *Hansard*, Parliament of Australia, Thursday, October 8, 1987, 961.
8 Nicholas Stern, "Stern Review on the Economics of Climate Change," HM Treasury, London, 2008, http://mudancasclimaticas.cptec.inpe.br/~rmclima/pdfs/destaques/sternreview_report_complete.pdf.
9 Marian Wilkinson, *The Fixer: The Untold Story of Graham Richardson* (Port Melbourne: William Heinemann), 1996, 420 pp.
10 Ibid., 282.
11 Ibid., 283.

12   Ibid., 288.
13   Wikipedia, "Joh Bjelke-Petersen," https://en.wikipedia.org/wiki/Joh_Bjelke-Petersen.
14   Wilkinson, *The Fixer*, 289.
15   Elizabeth May, "When Canada led the way: A short history of climate change," *Policy Options*, October 1, 2006, http://policyoptions.irpp.org/magazines/climate-change/when-canada-led-the-way-a-short-history-of-climate-change/.
16   H.L. Ferguson, "Twentieth Anniversary of the Toronto Conference on Our Changing Atmosphere: Implications for Global Security," *Canadian Meteorological and Oceanographic Society Bulletin* 36 (2008), 159–61.
17   Maria Taylor, *Global Warming and Climate Change: What Australia Knew and Buried* (Canberra: ANU Press, 2014), 24.
18   Marc Hudson, "25 years ago the Australian government promised deep emissions cuts, and yet here we still are," *The Conversation*, October 8, 2015, https://theconversation.com/25-years-ago-the-australian-government-promised-deep-emissions-cuts-and-yet-here-we-still-are-46805.
19   Taylor, *Global Warming*, 35.
20   R.J.L. Hawke, *Our Country Our Future: Statement on the Environment*, (Canberra: Australian Government Publishing Service, 1989), 66 pp., https://catalogue.nla.gov.au/catalog/2208246.
21   Ian Anderson, "Greening of industry: Problems for primary industries—Environmentalism is strong in Australia, but can the Hawke government balance the interests of industry, states and conservationists?" *New Scientist*, October 7, 1989, 10.
22   Taylor, *Global Warming*, 49.
23   *Hansard*, Parliament of Australia, Questions Without Notice, September 12, 1990, 1679.
24   Talberg et al., "Australian climate change policy."
25   *Hansard*, Parliament of Australia, Questions Without Notice, November 12, 1990, 3722.
26   A leadership spill—a term unique to Australian politics—"is a declaration that the leadership of a parliamentary body is vacant and thus open for re-election by the sitting members." A spill can be sparked when a rival to the existing head of the governing party calls for a leadership challenge. Australia had thirty-one spills between 2000 and 2015, and has become known as the "coup capital of the democratic world" (Wikipedia: https://en.wikipedia.org/wiki/Leadership_spill).
27   *Hansard*, Parliament of Australia, Questions Without Notice, September 12, 1990, 1679.

28 Clive Hamilton, *Scorcher: The Dirty Politics of Climate Change* (Melbourne: Schwartz, 2007), 159.
29 Guy Pearse, *High and Dry: John Howard, Climate Change and the Selling of Australia's Future* (London: Penguin, 2007), 465.
30 Soaring ocean temperatures in subsequent years have continued to fuel destructive bleaching of the Great Barrier Reef. The El Niño of 2023–24 has continued to damage the long-term viability of the reef complex. See, for example, https://www.cnn.com/2024/02/29/australia/australia-great-barrier-reef-bleaching-intl-hnk-scn/index.html.
31 Paul Pollard, "Missing the Target," Discussion Paper 51, Australia Institute, Canberra, 2003.
32 Pearse, *High and Dry*, 8.
33 Ibid., 9, 10.
34 Hamilton, *Scorcher*, 66 ff.
35 Pollard, "Missing the Target."
36 Clive Hamilton, "Australia hit its Kyoto target, but it was more a three-inch putt than a hole in one," *The Conversation*, July 15, 2015, https://www.theconversation.com/australia-hit-its-kyoto-target-but-it-was-more-a-three-inch-putt-than-a-hole-in-one-44831.
37 Hamilton, *Scorcher*, 66 ff.
38 Ibid.
39 Ibid., 97.
40 Talberg et al., "Australian climate change policy."
41 Pearse, *High and Dry*, 84.
42 Ibid., 87.
43 *Hansard*, Parliament of Australia, Questions Without Notice, June 5, 2002, 3163.
44 Pearse, *High and Dry*, 86.
45 *The Age*, "How big energy won the climate battle," June 5, 2005, https://www.theage.com.au/national/how-big-energy-won-the-climate-battle-20050730-geolrl.html.
46 Hamilton, *Scorcher*, 161.
47 Michelle Grattan, "Most voters want action on warming," *The Age*, November 7, 2006, https://www.theage.com.au/politics/victoria/most-voters-want-action-on-warming-20061107-ge3if6.html.
48 Lily Mayers and Lindy Kerin, "Climate change: Survey finds 77pc of Australians believe it is occurring," *ABC News*, September 25, 2016, http://www.abc.net.au/news/2016-09-26/77pc-of-australians-believe-climate-change-occurring-survey-says/7876416.
49 Pearse, *High and Dry*, 88.
50 Ibid., 90.

51 Ibid., 93.
52 National Emissions Trading Taskforce, *Possible Design for a National Greenhouse Gas Emissions Trading Scheme*, State and Territory Governments of Australia, 2006. Available at the National Library of Australia: https://catalogue.nla.gov.au/catalog/3748404.
53 Wikipedia, "Prime Ministerial Task Group on Emissions Trading," https://en.wikipedia.org/wiki/Prime_Ministerial_Task_Group_on_Emissions_Trading.
54 The secretary of a federal government department in Australia is equivalent to a deputy minister in Canada.

Chapter 5

1 Katharine Murphy, "Scott Morrison brings coal to question time: What fresh idiocy is this?" *Guardian*, February 9, 2017, https://www.theguardian.com/australia-news/2017/feb/09/scott-morrison-brings-coal-to-question-time-what-fresh-idiocy-is-this.
2 The right side of the political spectrum in Australia has been dominated for many years by a formal coalition between the centre-right Liberal Party and the more conservative National Party. This arrangement is a clone of right-wing politics in British Columbia, where the BC Liberal Party was liberal in name only and was, in essence, an informal and occasionally unwieldly coalition that brought both centre-right red conservatives and hardcore social conservatives into the same tent.
3 Australian Federal Election Speeches, Speech delivered in Brisbane, Queensland, November 14, 2007. Available at: http://electionspeeches.moadoph.gov.au/speeches/2007-kevin-rudd.
4 Katharine Murphy and Misha Schubert, "Rudd's decisive win", *The Age*, October 22, 2007, https://www.theage.com.au/politics/federal/rudds-decisive-win-20071022-ge63wh.html.
5 Interview with the author, March 7, 2017. Unless otherwise indicated, all quotations attributed to Mr. Rudd stem from that interview.
6 Australian Broadcasting Corporation, *Catalyst* (television series that explores "the forefront of science and technology"). Description at: http://www.abc.net.au/catalyst/stories/s948858.htm. Note: previously aired episodes from Season 3, including the climate-change episode, cannot be viewed outside Australia.
7 Garnaut, "Climate Change Review."
8 Michelle Grattan, "Turnbull and PM at loggerheads on Kyoto," *Sydney Morning Herald*, October 28, 2007, http://www.smh.com.au/news/federalelection2007environment/turnbull-pm-clash-over-kyoto/2007/10/27/1192941402381.html.

ENDNOTES

9   Philip Chubb, *Power Failure* (Collingwood, Australia: Black, 2014), Kindle version, location 356, Introduction, paragraph 20.

10 Julian Glover, "The lucky country?" *Guardian*, November 23, 2007, https://www.theguardian.com/environment/2007/nov/23/climatechange.australia.

11 Unlike in Canada, in Australian national elections candidates are preferentially ranked on the ballot. According to Wikipedia (https://en.wikipedia.org/wiki/Ranked_voting_system#cite_note-wsj1-9), "If no candidate is the first choice of more than half of the voters, then all votes cast for the candidate with the lowest number of first choices are redistributed to the remaining candidates based on who is ranked next on each ballot. If that does not result in any candidate receiving a majority, further rounds of redistribution occur ... until one candidate emerged with more than 50 per cent." Given that Australian politics are essentially dominated by two parties (Labor and a Liberal/National Party coalition), the popular vote result is termed the "two-party preferred vote."

12 The Senate of Australia comprises twelve members from each of the six states and two each from the Northern Territory and the Australian Capital Territory in which Canberra sits. Senators are elected for six years, but a system of rotation requires that half of the state members retire every three years. The four from the territories are elected at the same time as members of the House of Representatives; the duration of their terms of office coincides with that of the House.

13 Christopher Rootes, "The First Climate Change Election? The Australian General Election of 24 November 2007," *Environmental Politics* 17 (2008), 473, www.tandfonline.com/doi/full/10.1080/09644010802065815).

14 *Sydney Morning Herald*, "Australia ratifies Kyoto Protocol," December 4, 2007, https://www.smh.com.au/environment/australia-ratifies-kyoto-protocol-20071204-gdrqli.html.

15 Ross Garnaut, "Garnaut Climate Change Review Interim Report to the Commonwealth, State and Territory Governments of Australia," February 2008, http://www.garnautreview.org.au/CA25734E0016A131/WebObj/GarnautClimateChangeReviewInterimReport-Feb08/%24File/Garnaut%20Climate%20Change%20Review%20Interim%20Report%20-%20Feb%2008.pdf.

16 Commonwealth of Australia, "Carbon Pollution Reduction Scheme Green Paper," July 2008, http://pandora.nla.gov.au/pan/86984/20080718-1535/www.greenhouse.gov.au/greenpaper/report/pubs/greenpaper.pdf.

17 Australia, Department of Climate Change, "Carbon pollution reduction scheme: Green paper," 2008. Available from the Victoria Government Library Service: https://www.vgls.vic.gov.au/client/en_AU/vgls/search/detailnonmodal/ent:$002f$002fSD_ILS$002f0$002fSD_ILS:432149/

ada?qu=Greenhouses+---+Climate.&d=ent%3A%2F%2FSD_ILS%2F0%2FSD_ILS%3A432149%7EILS%7E56&ic=true&ps=300&h=8.

18 Carbon Market Watch, "Industry windfall profits from Europe's carbon market," March 2016, https://carbonmarketwatch.org/wp-content/uploads/2016/03/Policy-brief_Industry-windfall-profits-from-Europe%E2%80%99s_web_final-1.pdf.
19 Interview with the author, March 3, 2017. Unless otherwise indicated, all quotations attributed to Dr. Garnaut in this chapter stem from that interview.
20 Commonwealth of Australia, Carbon Pollution Reduction Scheme Bill 2009, http://parlinfo.aph.gov.au/parlInfo/download/legislation/bills/r4127_first/toc_pdf/09091b01.pdf;fileType=application%2Fpdf.
21 Australian Labor Party, "Climate change: The great moral challenge of our generation," YouTube, 2007, https://www.youtube.com/watch?v=CqZvpRjGtGM.
22 Emma Griffiths, "Restyled emissions scheme wins broad support," ABC Radio, May 4, 2009, https://www.abc.net.au/listen/programs/pm/restyled-emissions-scheme-wins-broad-support/1672686.
23 Julia Gillard, *My Story* (Sydney: Random House, 2014), 368.
24 Interview with the author, February 24, 2017. All quotations attributed to Dr. Steffen in this chapter stem from that interview.
25 Philip Coorey, "Malcolm Turnbull survives backbench revolt, for now," *Sydney Morning Herald*, November 25, 2009, https://www.smh.com.au/environment/climate-change/malcolm-turnbull-survives-backbench-revolt-for-now-20091124-jhea.html.
26 "Minchin a fruit loop say colleagues," *Australian*, November 12, 2009, http://www.theaustralian.com.au/archive/politics/minchin-a-fruit-loop-say-colleagues/news-story/c1b51c2fde11cc507665a6092be59e77.
27 Wikipedia, "Nick Minchin," https://en.wikipedia.org/wiki/Nick_Minchin.
28 Peter Martin and Michelle Grattan, "Abbott declares carbon tax 'toxic'," *Sydney Morning Herald*, March 28, 2011, https://www.smh.com.au/national/abbott-declares-carbon-tax-toxic-20110327-1cbwz.html.
29 *Sky News*, "Interview with Tony Abbott," YouTube, 2009, https://www.youtube.com/watch?v=i5QXblcJAr4&t=9m18s.
30 *ABC News*, "Senior Liberals desert Turnbull," November 25, 2009, http://www.abc.net.au/news/2009-11-26/senior-liberals-desert-turnbull/1158164.
31 Interview with the author, March 15, 2017. Except where otherwise indicated, all quotations from Mr. Combet in this book stem from that interview.
32 David Marr, "Political Animal: The Making of Tony Abbott," *Quarterly Essay* 47 (2012), Kindle edition, location 1194 (Chapter title: "Mr. Lucky," paragraph 18).
33 *Hansard*, Senate of Australia, November 17, 2009, 7965.
34 Interview with the author, March 21, 2017. Cathy Alexander was with Associated Press in the press gallery in 2009 and was later a PhD candidate at the University

of Melbourne. All quotations attributed to Dr. Alexander in this book stem from that interview.

35 *Hansard*, Senate of Australia, November 18, 2009, 8153.
36 *Hansard*, Senate of Australia, November 30, 2009, 9285.
37 Ibid., 9288.
38 Ibid., 9291.
39 Ibid., 9401.
40 Ibid., 9430.
41 Ibid., 9582.
42 Ibid., 9583.
43 Ibid., 9544
44 Interview with the author; anonymity requested.
45 Meraiah Foley, "Australia's Rudd Looks for Success in Copenhagen," *New York Times*, December 14, 2009, http://www.nytimes.com/2009/12/15/science/earth/15australiaclimate.html.
46 Gillard, *My Story*, 371.
47 Interview with the author, March 24, 2048.
48 *Hansard*, House of Representatives of Australia, February 8, 2010, 570.
49 Ibid., 578–84.
50 *Hansard*, Senate of Australia, February 22, 2010, 698.
51 Chubb, *Power Failure*, location 1574, p. 114.
52 Ibid., locations 1639–40, p. 113.
53 See, for example, Christopher Rootes, "A referendum on the carbon tax? The 2013 Australian election, the Greens and the environment," *Environmental Politics* 23, no. 1 (2014): 166–73, DOI:10.1080/09644016.2014.878088, ISSN 0964-4016 print/ISSN 1743-8934 online.
54 Chubb, *Power Failure*, location 519, p. 114.
55 Ibid., location 1547, p. 112.
56 Gillard, *My Story*, 378.
57 Parliament of Australia, "Transcript of doorstop interview: 27 April 2010: Nepean Hospital, Penrith," April 27, 2020, http://parlinfo.aph.gov.au/parlInfo/search/display/display.w3p;query=Id%3A%22media%2Fpressrel%2FL9JW6%22.
58 Chubb, *Power Failure*, location 1777, p. 133.
59 Paul Kelly, "How to design and deliver reform that makes a real difference: What recent history has taught us as a nation," in eds. E. Lindquist, S. Vincent and J. Wanna, *Delivering Policy Reform: Anchoring Significant Reforms in Turbulent Times* (ANU Press, 2011), 43–52. Retrieved from: http://www.jstor.org/stable/j.ctt24h91v.8.

Chapter 6

1. "Australian coup had little grâce," *Globe and Mail*, June 27, 2010, https://www.theglobeandmail.com/opinion/editorials/australian-coup-had-little-grace/article4323044/.
2. Greg Ansley, "Game on: Gillard set for a fight," *New Zealand Herald*, section B, June 25, 2010, https://www.nzherald.co.nz/world/game-on-gillard-set-for-a-fight/CZIWXFKHRHCGA3SFMFQVE3U67Q/.
3. Gillard, *My Story*, 380.
4. Gillard's statement can be seen in a brief video clip archived at: https://marchudson.net/academia/phd-2014-to-2017/australia/no-carbon-tax-under-the-government-i-lead/.
5. Gillard, *My Story*, 385.
6. Chubb, *Power Failure*, location 2200, p. 175.
7. Ibid., location 2205, p. 176.
8. Gillard, *My Story*, 386.
9. Ibid., 389.
10. *Hansard*, Parliament of Australia, February 24, 2011, 1418.
11. Video available at: https://www.youtube.com/watch?v=9UTHsfTEZ3w. Gillard's concession to the word *tax* is at 4:22.
12. Chubb, *Power Failure*, location 2329, p. 187.
13. Ibid., location 2357, p. 190.
14. Parliament of Australia, Clean Energy Bill 2011, http://parlinfo.aph.gov.au/parlInfo/search/display/display.w3p;query=Id:legislation%2Fbillhome%2Fr4653.
15. Gillard, *My Story*, 388.
16. Parliament of Australia, Explanatory Memorandum, Clean Energy Bill 2011, 14.
17. Greg Combet, *The Fights of My Life* (Carlton: Melbourne University Press, 2014), 261.
18. Ibid., 262.
19. A description of the Murdoch group published in July 2011: Ben Eltham, "A close look at Abbott's Direct Action plan," ABC, July 27, 2011, https://www.abc.net.au/news/2011-07-21/eltham-a-close-look-at-abbott-direct-action-plan/2804206.
20. Data available at: https://www.google.ca/publicdata/explore?ds=d5bncppjof8f9_&met_y=en_atm_co2e_pc&idim=country:AUS:NZL:CAN&hl=en&dl=en.
21. Combet, *Fights of My Life*, 274.
22. Chubb, *Power Failure*, location 2750, p. 224.
23. UNSW Community, "Dr Anne Summers AO: Her Rights at Work: The political persecution of Australia's first female PM," YouTube, August 31, 2012, https://www.youtube.com/watch?v=R3pvCr8Msyo.

ENDNOTES

24 Dennis Shanahan, "Another crushing blow for Gillard," *Australian*, June 24, 2013, http://www.theaustralian.com.au/opinion/columnists/poll-delivers-another-crushing-blow-for-julia-gillard/news-story/4a64ce9cb64c4cbe6dd33f97ae0 6aa70. In his "crushing defeat" comment, Shanahan was referring to the federal election that had to be held later that year. In Australia, MPs in the House of Representatives are elected for a maximum term of three years, and senators for six years. Half of the Senate seats are contested in each federal election (except in the case of a double dissolution).

25 BBC News, "Australia politics: Gillard, Rudd in leadership vote," June 26, 2013, http://www.bbc.com/news/world-asia-23058602.

26 BBC News, "Australia's Kevin Rudd in party leadership vow," March 22. 2013, http://www.bbc.com/news/world-asia-21891382.

27 Jeffrey Simpson. "Another very Australian coup," *Globe and Mail*, June 28, 2013, accessed December 1, 2017.

28 Parliament of Australia, "Transcript of joint press conference Townsville," July 16 2013, http://parlinfo.aph.gov.au/parlInfo/download/media/pressrel/2597365/upload_binary/2597365.pdf.

29 Commonwealth of Australia, "How Australia's carbon price is working one year on," Department of Industry, Innovation, Climate Change, Science, Research and Tertiary Education, ISBN: 978-1-925006-24-7 (print) 978-1-925006-25-4 (web), https://parlinfo.aph.gov.au/parlInfo/search/display/display.w3p;query=Id:%22library/lcatalog/00701944%22.

30 Alex Frankel, "From 'axe the tax' to 'climate consensus': How Abbott reshaped our climate story," *Guardian*, April 25, 2015, https://www.theguardian.com/commentisfree/2015/apr/28/from-axe-the-tax-to-climate-consensus-how-abbott-reshaped-our-climate-story.

31 Interview with the author, February 19, 2017.

32 Xavier Smerdon, "Acclimatising to the top job," *Probono*, January 23, 2015, https://probonoaustralia.com.au/news/2015/01/acclimatising-to-the-top-job/.

33 Emma Griffiths, "Tony Abbott introduces legislation to repeal carbon tax after 'electricity bill' row," ABC News, November 12, 2013, https://www.abc.net.au/news/2013-11-13/abbott-introduces-carbon-tax-repeal-bill/5088524.

34 *Hansard*, Commonwealth of Australia, House of Representatives, November 13, 2013, 75.

35 Ibid., 76.

36 Ibid., 77–78.

37 Rintoul, S. (December 12, 2009), "The town that turned up the temperature," *The Weekend Australian*, retrieved May 26, 2014, from http://www.theaustralian.com.au/archive/politics/the-town-that-turnedup-the-temperature/story-e6frgczf-1225809567009.

38   *Insiders*, Australian Broadcasting Corporation, September 1, 2013, https://www.abc.net.au/insiders/tony-abbott-joins-insiders/4927470.
39   Australian Associated Press, "Turnbull blasts Abbott's 'recklessness'," *Sydney Morning Herald*, February 8, 2010, https://www.smh.com.au/environment/climate-change/turnbull-blasts-abbotts-recklessness-20100208-nmb0.html.
40   Eltham, "A close look," https://www.abc.net.au/news/2011-07-21/eltham-a-close-look-at-abbott-direct-action-plan/2804206.
41   Malcolm Turnbull, "Abbott's climate change policy is bullshit," *Sydney Morning Herald*, December 7, 2009, https://www.smh.com.au/politics/federal/abbotts-climate-change-policy-is-bullshit-20091207-kdmb.html.
42   Interview with the author, February 24, 2017.
43   *Hansard*, Senate of Australia, July 17, 2014, 5247.
44   *Lateline*, "Shut-up: Clive Palmer challenges climate report findings," YouTube, April 3, 2014, https://www.youtube.com/watch?v=bh_BNKpt9jg.
45   BBC News, "Australia saw hottest year on record in 2013," January 3, 2014, https://www.bbc.com/news/world-asia-25573712.
46   Video of the full press conference is available at: https://www.youtube.com/watch?v=mNdnA4QfuF8.
47   Heath Aston, "Al Gore told to avoid Clive Palmer's environment circus," *Sydney Morning Herald*, July 2, 2014, https://www.smh.com.au/politics/federal/al-gore-told-to-avoid-clive-palmers-environment-circus-20140701-3b6l2.html.
48   Andrew Critchlow, "Australia abandons disastrous green tax on emissions," *Telegraph*, July 17, 2014, https://www.telegraph.co.uk/finance/commodities/10972902/Australia-abandons-disastrous-green-tax-on-emissions.html.
49   "Australia abolishes tax on carbon emissions," *Financial Times*, July 17, 2014, https://www.ft.com/content/d852822a-0d67-11e4-bcb2-00144feabdc0.
50   James Whitmore et al., "Carbon tax repealed: experts respond," *The Conversation*, July 16 2014, https://theconversation.com/carbon-tax-repealed-experts-respond-29154.
51   Julia Baird, "A Carbon Tax's Ignoble End," *New York Times*, July 24, 2014, https://www.nytimes.com/2014/07/25/opinion/julia-baird-why-tony-abbott-axed-australias-carbon-tax.html.
52   Steven Geroe, "Addressing climate change through a low-cost, high-impact carbon tax," *Journal of Environment and Development* 28, no. 1 (2019): 3–27.
53   Australian Associated Press, "Poll gives Tony Abbott record low approval rating," *Guardian*, February 15, 2015, https://www.theguardian.com/australia-news/2015/feb/09/poll-tony-abbott-record-low-approval-rating.

Chapter 7

1. Richard Littlemore, "Jaccard analysis blunts NDP's carbon tax axe," *DeSmog*, April 11, 2009, https://www.desmogblog.com/jaccard-analysis-blunts-ndps-carbon-tax-axe.
2. *Hansard*, Legislative Assembly of British Columbia, May 29, 2008, 13142, https://www.leg.bc.ca/content/Hansard/38th4th/H0529pm-05.pdf.
3. Justine Hunter, "NDP Vote Against Carbon Tax," *Globe and Mail*, May 30, 2008, https://www.theglobeandmail.com/news/national/ndp-vote-against-carbon-tax/article959518/.
4. Littlemore, "Jaccard analysis," https://www.desmogblog.com/jaccard-analysis-blunts-ndps-carbon-tax-axe.
5. Quoted by *DeSmog*'s Richard Littlemore on April 17, 2008, in "Tzeporah Berman leading climate policy charge," https://www.desmogblog.com/tzeporah-berman-leading-climate-policy-charge.
6. Jonathan Fowlie, "As fuel costs soar, NDP steps up campaign against carbon tax," *Vancouver Sun*, June 17, 2008, A3.
7. Excerpt from Ms. James's speech of June 17, 2008, posted on: www.bcndpcaucus.ca. Accessed via the Wayback Machine, February 16, 2021: https://web.archive.org/web/20080802001547/http://www.bcndpcaucus.ca/en/james_launches_campaign_to_axe_the_gas_tax.
8. *CBC News*, "B.C. premier says NDP plan to 'axe the tax' is playing politics," June 17, 2008, https://www.cbc.ca/news/canada/british-columbia/b-c-premier-says-ndp-plan-to-axe-the-tax-is-playing-politics-1.760686.
9. Andrew MacLeod, "The Carbon Campaign of '09", *The Tyee*, June 5, 2008, https://thetyee.ca/News/2008/06/05/CarbonCampaign/.
10. The petition is available via the Wayback Machine at: https://web.archive.org/web/20080701072836/http://www.bcndpcaucus.ca/en/node/2104.
11. Michael Smyth, "Honk if you want carbon tax gassed," *Province*, June 17, 2008, A4.
12. Interview with the author, May 1, 2017. Unless otherwise indicated, all quotations attributed to Dr. Olewiler in this book stem from that interview.
13. Interview with the author, December 15, 2022. All quotations in this book attributed to Dr. Green stem from that interview.
14. David Green, "Why 70 Economists Urge BC Carbon Tax," *The Tyee*, November 1, 2007, https://thetyee.ca/Views/2007/11/01/CarbonTax/?commentsfilter=0.
15. Interview with the author, October 23, 2017. Unless otherwise indicated, all quotations in this book attributed to Ms. Berman stem from that interview.
16. Lindsay Kines, "Carbon tax turns up the heat on NDP chief," *Victoria Times-Colonist*, June 25, 2008, A3.

17. Richard Littlemore, "Bill Tieleman, the NDP Svengali?" *BC Business Magazine*, February 7, 2011.
18. MacLeod, "Carbon Campaign," https://thetyee.ca/News/2008/06/05/CarbonCampaign/.
19. Description available at: https://billtieleman.blogspot.com/2008/06/axe-bc-gas-tax-protest-on-facebook.html.
20. Interview with the author, July 12, 2017. Unless otherwise indicated, all quotations attributed to Mr. Tieleman in this book stem from that interview.
21. Littlemore, "Bill Tieleman."
22. D. Simon Jackson, "Will the BC election prove the naysayers wrong?" *CBC News*, May 4, 2009, https://www.cbc.ca/news/canada/will-the-b-c-election-prove-the-naysayers-wrong-1.817303.
23. Ian Bailey, "Suzuki decries NDP plan to axe carbon tax," *Globe and Mail*, April 18, 2009, https://www.theglobeandmail.com/news/politics/suzuki-decries-ndp-over-plan-to-axe-carbon-tax/article4193359/?service=print.
24. CHEK-TV News, "2009 04 16 CH carbon tax," YouTube, April 13, 2009, https://www.youtube.com/watch?v=zCt6YGmuHmw.
25. Ibid.
26. CTV News, "NDP promises to axe B.C. carbon tax," September 25, 2008, https://bc.ctvnews.ca/ndp-promises-to-axe-b-c-carbon-tax-1.328022.
27. *Hansard*, Hansard, Commonwealth of Australia, House of Representatives, November 13, 2013, 76.
28. Pembina Institute, "B.C. Climate Change Policy: Media Backgrounder," April 13, 2009, https://www.pembina.org/pub/1810.
29. A copy of the letter was provided to the author by Ms. Berman.
30. Doug Ward, "Key supporter quits NDP over carbon tax," *Vancouver Sun*, April 17, 2009, A1.
31. CTV News, "NDP promises," September 25, 2008, https://bc.ctvnews.ca/ndp-promises-to-axe-b-c-carbon-tax-1.328022.
32. Gary Mason, "B.C.'s New Democrats picked the wrong fight with the carbon tax," *Globe and Mail*, May 12, 2009.
33. Email communications with the author, January 9 and 16, 2023.
34. Kathryn Harrison, "A tale of two taxes: The fate of environmental tax reform in Canada," *Review of Policy Research* 29 (2012), 383–407.
35. Interview with the author, March 4, 2021. Unless otherwise indicated, all quotations attributed to Mr. Chandra Herbert in this book stem from that interview.
36. Vaughn Palmer, "James makes some hard choices, talks tough with shadow cabinet," *Vancouver Sun*, June 12, 2009, A3.
37. "Carbon tax struggle ends," *Victoria Times-Colonist*, June 13, 2009, 14.

38 Richard Littlemore, "BC NDP Leader Accepts BC Carbon Tax (Bravo! Carole James)," *DeSmog*, June 11, 2009, https://www.desmogblog.com/bc-ndp-leader-accepts-bc-carbon-tax-bravo-carole-james.

39 Will McMartin, "Making Sense of a Brutal Political Season in BC," *The Tyee*, December 10, 2010, https://thetyee.ca/Opinion/2010/12/10/BrutalSeason/.

40 Doug Ward, "No third chance for Carole James," *Vancouver Sun*, December 7, 2010, A2.

41 Littlemore, "Bill Tieleman."

42 Ian Austin, "Backroom boys forced James out," *Province*, December 8, 2010, A6.

Chapter 8

1 Fiona Anderson, "You'll pay more for haircuts, meals and school supplies," *Vancouver Sun*, July 24, 2009, A1.

2 Vaughn Palmer, "Liberals sing different tune on HST after Ottawa dangles carrot," *Vancouver Sun*, July 24, 2009, A3.

3 Steve Burgess, "Nine Per Cent Gordo," *The Tyee*, October 19 2010, https://thetyee.ca/Opinion/2010/10/19/NinePerCentGordo/.

4 Jonathan Fowlie, "It's not always popular to do what you believe in your heart is right," *Vancouver Sun*, November 4, 2010, A1.

5 Video of Margaret Thatcher's speech can be found at: https://www.c-span.org/video/?9846-1/global-environmental-issues.

6 Michael Smyth, "Voting Rules a Big Relief for Abbott," *Province*, February 13, 2011, A3.

7 See, for example: Brian Hutchinson, "Documents raise questions about BC Liberals' change of course on privatizing liquor board," *National Post*, July 19, 2012; Mark Hume and Justine Hunter, "MLA calls for investigation of Clark over sale of BC Rail," *Globe and Mail*, November 7, 2012; Armina Ligaya, "Ethnic vote scandal 'dashes' BC Liberals' May election hopes," *National Post*, March 3, 2013.

8 Interview with the author, May 28, 2021. Unless otherwise indicated, all quotations attributed to Ms. Clark in this book stem from that interview.

9 Kinder Morgan Inc. had purchased the Trans Mountain Pipeline in 2005.

10 Interview with the author, December 13, 2022. Unless otherwise indicated, all quotations from Mr. Lake in this book stem from that interview.

11 The conditions are listed in a Government of BC Information Bulletin available at: https://news.gov.bc.ca/releases/2012ENV0049-001120.

12 Federal Reserve Bank of St. Louis, "Global price of LNG, Asia," https://fred.stlouisfed.org/series/PNGASJPUSDM.

13 Government of British Columbia, "Budget and Fiscal Plan, 2013/14–2015-16," February 19, 2013, https://www.bcbudget.gov.bc.ca/2013/bfp/2013_budget_fiscal_plan.pdf.

14 Government of British Columbia, "New British Columbia Prosperity Fund Will Ensure Lasting Benefits," February 12, 2013, https://news.gov.bc.ca/releases/2013PREM0018-000231#:~:text=The%20purpose%20of%20the%20fund,through%20the%20BC%20Prosperity%20Fund.

15 Jonathan Fowlie, "Did the BC Liberal party scoop government on the carbon tax?" *Vancouver Sun*, April 3, 2013, https://vancouversun.com/news/staff-blogs/did-the-bc-liberal-party-scoop-government-on-the-carbon-tax.

16 Ian Bailey, "B.C.'s Clark vows to freeze carbon tax for five years," *Globe and Mail*, April 3, 2013, https://www.theglobeandmail.com/news/british-columbia/bcs-clark-vows-to-freeze-carbon-tax-for-five-years/article10728482/.

17 Interview with the author, December 21, 2022. Unless otherwise indicated, all quotations attributed to Mr. Rhone in this book stem from that interview.

18 Interview with the author, December 20, 2022. Unless otherwise indicated, all quotations attributed to Ms. Polak in this book stem from that interview.

19 CBC News, "Carbon tax freeze part of B.C. Liberal election pledge," April 4, 2013, https://www.cbc.ca/news/canada/british-columbia/carbon-tax-freeze-part-of-b-c-liberal-election-pledge-1.1389584.

20 Bailey, "B.C.'s Clark," https://www.theglobeandmail.com/news/british-columbia/bcs-clark-vows-to-freeze-carbon-tax-for-five-years/article10728482/.

21 Michael Smyth, "Clark played her strongest card all along," *Province*, May 15, 2013, A3.

22 *Hansard*, British Columbia, 40th Parliament, vol. 1, no. 6, July 3, 2013, 97 ff.

23 Natural Gas Intelligence, "Tax Advantage Sought for Canadian LNG Projects," February 18, 2013, https://www.naturalgasintel.com/tax-advantage-sought-for-canadian-lng-projects/.

24 Government of British Columbia, "Budget Speech 2014," p. 9, https://www.bcbudget.gov.bc.ca/2014/speech/2014_Budget_Speech.pdf.

25 Rob Kreklewetz and John Bassindale, "Overview of BC's Liquefied Natural Gas (LNG) Tax Regime," *Tax & Trade Blog*, May 13, 2015, https://www.taxandtradelaw.com/Tax-Trade-Blog/overview-of-bc-s-liquefied-natural-gas-lng-tax-regime.html.

26 Julie Gordon, "British Columbia sets new LNG income tax at 3.5 percent," Reuters, October 21, 2014, https://www.reuters.com/article/canada-us-canada-lng-tax-idCAKCN0IA2NL20141021.

27 Vaughn Palmer, "On LNG promises, disclaimers tell a tale," *Vancouver Sun*, February 28, 2014, A12.

28 Ms. Polak served as minister of the environment from June 2013 to June 2017.

29 Federal Reserve Bank of St. Louis, "Asia," https://fred.stlouisfed.org/series/PNGASJPUSDM

30 Interview with the author, July 9, 2021. Unless otherwise indicated, all quotations from Ms. Smith in this book stem from that interview.

31 Government of British Columbia, "B.C. names Climate Leadership Team," May 12, 2015, https://news.gov.bc.ca/releases/2015PREM0032-000655.

32 CLIMATE LEADERSHIP TEAM MEMBERS:
Mike Bernier, chair, MLA for Peace River South
Jordan Sturdy, deputy chair, MLA for West Vancouver-Sea to Sky
Susanna Laaksonen-Craig, head, Climate Action Secretariat
Tom Pedersen, executive director, Pacific Institute for Climate Solutions, University of Victoria
Nancy Olewiler, economist, School of Public Policy, Simon Fraser University
James Tansey, associate professor, UBC Sauder School of Business, and CEO and co-founder of Offsetters
Paul Ives, mayor of the Town of Comox
Luke Strimbold, mayor of the Village of Burns Lake
Linda Hepner, mayor of the City of Surrey
James Gorman, president and CEO of the Council of Forest Industries
Greg McDougall, CEO, Harbour Air
Tim Newton, former vice president of strategic planning with Powerex, and director of Columbia Power
David Keane, president, BC LNG Alliance
Merran Smith, executive director, Clean Energy Canada
Matt Horne, associate regional director for British Columbia, Pembina Institute
Tzeporah Berman, former co-director of Greenpeace International's Global Climate and Energy Program
Chief Ian Campbell, Squamish First Nation
Chief Zach Parker, Ulkatcho First Nation
Chief Michelle Edwards, Cayoose Creek Indian Band

33 Climate Leadership Team, "Recommendations to Government," October 31, 2015, https://www2.gov.bc.ca/assets/gov/environment/climate-change/action/clp/clt-recommendations-to-government_final.pdf.

34 That community fundamentally included the natural gas industry, which was fearful of the impact a rising carbon tax might have on its profitability. During the CLT deliberations, that point was made unreservedly by David Keane, the president of the BC LNG Alliance and a member of the team. Given the concerns of the sector he represented, Keane was unable to support the carbon pricing recommendation in the CLT report. He was the only member to object.

ENDNOTES

35   The letter is available at: https://cleanenergycanada.org/wp-content/uploads/2016/05/Climate-action-letter-to-Premier-Clark-from-CLT-Members-May-16.pdf.

36   Government of British Columbia, "B.C.'s Climate Leadership Plan to cut emissions while growing the economy," August 19, 2016, https://news.gov.bc.ca/releases/2016PREM0089-001501.

37   CBC News, "B.C. Environment Minister Mary Polak defends provincial climate change plan," August 22, 2016, https://www.cbc.ca/news/canada/british-columbia/environment-minister-climate-plan-1.3731243.

38   Graham Thomson, "No province is really a leader on climate change," *Edmonton Journal*, August 22, 2016, https://edmontonjournal.com/news/politics/graham-thomson-no-province-is-really-a-leader-on-climate-change.

39   Lien Yeung and Jason Proctor, "B.C. Premier Christy Clark's climate change plan does not raise carbon tax," *CBC News*, August 19, 2016, https://www.cbc.ca/news/canada/british-columbia/b-c-premier-christy-clark-s-climate-change-plan-does-not-raise-carbon-tax-1.3728317.

40   Lisa Johnson, "B.C.'s delayed Climate Leadership Plan expected today," *CBC News*, August 19, 2016, https://www.cbc.ca/news/canada/british-columbia/b-c-s-delayed-climate-leadership-plan-expected-today-1.3726795.

41   Canadian Press, "B.C. climate plan needs carbon tax hikes, major policy changes, experts say," *Nanaimo News Now*, August 18, 2016, https://nanaimonewsnow.com/2016/08/19/b-c-climate-plan-needs-carbon-tax-hikes-major-policy-changes-experts-say/.

42   *CBC News*, "B.C. Environment Minister Mary Polak defends provincial climate change plan," August 22, 2016. Available at: https://www.cbc.ca/news/canada/british-columbia/environment-minister-climate-plan-1.3731243.

43   Carol Linnitt, "Christy Clark's Secret Consultations with Oil and Gas Donors Revealed as B.C. Introduces Bill to Ban Big Money in Politics," *The Narwhal*, September 18, 2017, https://thenarwhal.ca/christy-clark-s-secret-consultations-oil-and-gas-donors-revealed-b-c-introduces-bill-ban-big-money-politics/.

44   Nicolas Graham, Shannon Daub and Bill Carroll, "Mapping political influence: Political donations and lobbying by the fossil fuel industry in BC," Canadian Centre for Policy Alternatives, March 2017, 4 pp.

45   Nancy Macdonald, "The B.C. election delivered Christy Clark a hollow victory," *Maclean's*, May 10, 2017, https://www.macleans.ca/politics/the-b-c-election-delivered-christy-clark-a-hollow-victory/.

46   Dirk Meissner, "B.C. premier says 5 conditions met on Trans Mountain pipeline," *CTV News*, January 11, 2017, https://bc.ctvnews.ca/b-c-premier-says-5-conditions-met-on-trans-mountain-pipeline-1.3237261.

47 Chris Genovali, "Opinion: Christy Clark's 'five conditions con'," *Vancouver Sun*, February 9, 2017, https://vancouversun.com/opinion/opinion-christy-clarks-five-conditions-con.

48 Martyn Brown, "The irony in Christy Clark's demise," *Georgia Straight*, June 4, 2017.

49 See pages 67–68 of the BC Budget Update of September 2017: https://www.bcbudget.gov.bc.ca/2017_Sept_Update/bfp/2017_Sept_Update_Budget_and_Fiscal_Plan.pdf.

Chapter 9

1 Bob Inglis and Arthur B. Laffer, "An emissions plan conservatives could warm to." *New York Times*, December 8, 2008, https://www.nytimes.com/2008/12/28/opinion/28inglis.html.

2 John F. Kennedy Presidential Library and Museum, "Former U.S. Congressman Bob Inglis to receive JFK Profile in Courage Award for stance on climate change," April 13, 2015, https://www.jfklibrary.org/about-us/news-and-press/press-releases/2015-profile-in-courage-announcement.

3 Paul Krugman, "Nobody Cares About Biden's Energy Policy. Great!" *New York Times*, November 15, 2022, https://www.nytimes.com/2022/11/15/opinion/biden-climate-change-ira.html.

4 But it is a tax with one major weakness: it was not a tax shift—revenue-neutrality was not a core requirement. Instead of reducing other taxes, revenues were directed toward sustainability issues, which makes the tax regrettably vulnerable to the vagaries of national politics. At the time, however, the carbon tax was a part of a broad set of tax reforms; those lowered its visibility and made it initially less vulnerable to opposition, according to a number of sources, including: Darragh Conway et al, *Tipping the Balance: Lessons on Building Support for Carbon Prices* (Berlin/Amsterdam/Freiburg: Adelphi, Climate Focus B.V., Perspectives Climate Group, 2019), https://www.adelphi.de/en/publication/tipping-balance.

Chapter 10

1 See: World Economic Forum, "Climate change is costing the world $16 million per hour," October 12, 2023, https://www.weforum.org/agenda/2023/10/climate-loss-and-damage-cost-16-million-per-hour/. This report draws directly on a comprehensive analysis published in 2023 in the top-drawer journal *Nature Communications*, entitled "The global costs of extreme weather that are attributable to climate change," available at https://doi.org/10.1038/s41467-023-41888-1.

ENDNOTES

2   Government of Canada, "Government of Canada Announces Pan-Canadian Pricing on Carbon Pollution," October 3, 2016, https://www.canada.ca/en/environment-climate-change/news/2016/10/government-canada-announces-canadian-pricing-carbon-pollution.html.

3   Government of Canada, Greenhouse Gas Pollution Pricing Act, https://laws-lois.justice.gc.ca/eng/acts/g-11.55/.

4   See: Government of Alberta, "Technology Innovation and Emissions Reduction System," https://www.alberta.ca/technology-innovation-and-emissions-reduction-system; and "TIER regulation Fact Sheet," https://www.alberta.ca/system/files/custom_downloaded_images/ep-fact-sheet-tier-regulation.pdf.

5   A full description can be found at: Government of Canada, "Output-Based Pricing System," https://www.canada.ca/en/environment-climate-change/services/climate-change/pricing-pollution-how-it-will-work/output-based-pricing-system.html.

6   Dale Beugin et al., "Which Canadian climate policies will have the biggest impact by 2030?" *440 Megatonnes* (Canadian Climate Institute), March 21, 2024, https://440megatonnes.ca/insight/industrial-carbon-pricing-systems-driver-emissions-reductions/.

7   A necessary caveat: It *was* sacrosanct in British Columbia, beginning in 2008. But in the latter years of Christy Clark's government—and following the lifting of her carbon tax freeze in 2018 by the newly elected left-wing government of John Horgan—some of the annually increasing C tax revenue was diverted into the provincial treasury pool, rather than being applied to *further* income tax relief. As the *Globe and Mail* editorial board noted on February 15, 2023, "tangential tax breaks such as children's fitness and arts credits, and existing subsidies to the film and TV industry," all to be supported by carbon tax revenues, were included in the last budget of the BC Liberals in 2017. Months later, in its "Budget and Fiscal Plan–Budget 2017 September Update," the Horgan government repealed Part 2 of the Carbon Tax Act of 2008, a move that erased the requirement "that revenue measures be introduced to offset carbon tax revenues." From that moment on, pure revenue neutrality, as originally required by the tax-shifting philosophy of the Gordon Campbell government in 2008, was no longer fully in place in BC. The distribution of C tax revenue in British Columbia does continue to embrace fairness, however: supports for low- and middle-income families in the province have continued to increase, and for a family of four with a net income of $50,170 or less in 2024, $893.50 in direct cash subventions will be provided, up from $700 in 2018. Revenues will also be used to support a range of other programs, including installation of EV chargers, primarily in the north, as well as rebates for heat pump installation and the like. But in an ideal world, restoring full revenue neutrality by reducing other taxes would be a politically

wiser pathway. In not doing so, British Columbia's carbon tax has been made more politically vulnerable today to the chants of nefarious axe-the-tax actors than it was in 2008.

8   The remaining 10 per cent supports climate-action programs, subventions to businesses, and the provision of supports targeted to unique circumstances including food security challenges in Canada's far north.

9   Government of Canada, "Canada Carbon Rebate amounts for 2024–25," https://www.canada.ca/en/department-finance/news/2024/02/canada-carbon-rebate-amounts-for-2024-25.html.

10   Robson Fletcher, "If Canada axed its carbon tax—and rebates—this is how different households would gain or lose," *CBC News*, December 5, 2023, https://www.cbc.ca/news/canada/calgary/axe-the-tax-and-carbon-rebate-how-canada-households-affected-1.7046905.

11   Radio Canada International, "Millions of Canadians get their carbon tax rebates today. So why do many not believe it?" January 15, 2024, https://ici.radio-canada.ca/rci/en/news/2041709/millions-of-canadians-get-their-carbon-tax-rebates-today-so-why-do-many-not-believe-it.

12   Rachel Aiello, "Liberals rebrand Canada's carbon tax rebate," *CTV News*, February 14, 2024, https://www.ctvnews.ca/politics/canada-s-carbon-tax-rebate-system-has-been-rebranded-policy-unchanged-1.6768770.

13   "Pierre Poilievre makes journalists the target," *Toronto Star*, March 2, 2024, https://www.thestar.com/opinion/editorials/pierre-poilievre-makes-journalists-the-target/article_cf206868-d5ab-11ee-8139-d3bd6700ef5b.html.

14   Benjamin Shingler, "What's behind the carbon tax, and does it work?" *CBC News*, March 22, 2024, https://www.cbc.ca/news/climate/carbon-tax-controversy-1.7151551.

15   Video clip of Canadian Defence Minister Bill Blair speaking in the House on March 17, 2024: https://www.youtube.com/watch?v=W4HaL74ujDU.

16   This conclusion is robustly supported by a recent comprehensive economic analysis of the impact of revenue-neutral and non-revenue-neutral carbon taxation schemes in place for up to three decades in Europe and Canada. The paper, published in the *Journal of the European Economic Association* in late 2023, asks a simple question: Does putting a price on carbon in fact lead to "greenflation"? The paper concludes: "Controlling for country and year fixed effects, as well as economic controls, we find no robust evidence of an inflationary response, on average. This result holds both for European countries and Canadian provinces, and survives a battery of robustness checks." Full results can be found at: https://academic.oup.com/jeea/article/21/6/2518/7079134.

ENDNOTES

17 Jason Markusoff, "There's now a Bank of Canada number for carbon tax's impact on inflation. It's small," *CBC News*, September 8, 2023, https://www.cbc.ca/news/canada/calgary/carbon-tax-inflation-tiff-macklem-calgary-1.6960189.

18 Ismail Shakil and Nia Williams, "Canada to pause carbon tax on home heating oil for three years," Reuters, October 27, 2023, https://www.reuters.com/sustainability/cop/canada-pause-carbon-tax-home-heating-oil-three-years-2023-10-26/.

19 Halifax, NS, Average Retail Price for Household Heating Fuel, Y Charts. Available at: https://ycharts.com/indicators/halifax_ns_average_retail_price_for_household_heating_fuel.

20 See: Conservative Party of Canada, website, https://www.conservative.ca/?gclid=CjwKCAjwnv-vBhBdEiwABCYQA28Vj_Y--0W2xeV5FIes0QZ5iWrSanFrJIW6le_j_BNYChofQacgIhoCJCQQAvD_BwE.

21 See: Secure the Environment: The Conservative Plan to Combat Climate Change, https://d3n8a8pro7vhmx.cloudfront.net/marilyngladu/pages/65/attachments/original/1618503903/CPC-EnvironmentPolicy-EN-WEB-FINAL.pdf?1618503903.

22 Government of Canada, Standing Committee on Government Operations and Estimates, Meeting 113, March 27, 2024. Proceedings available at: https://www.ourcommons.ca/Content/Committee/441/OGGO/Evidence/EV12987694/OGGOEV113-E.PDF.

23 Government of Canada, Standing Committee on Government Operations and Estimates, Meeting 114, March 28, 2024. Proceedings available at: https://parlvu.parl.gc.ca/Harmony/en/PowerBrowser/PowerBrowserV2/20240331/-1/41275.

24 "At Issue," *The National* (*CBC News*), March 28, 2024.

25 *Global News*, "Carbon tax hike: Premiers Danielle Smith, Blaine Higgs pressed on alternative climate plan," March 28, 2024, https://globalnews.ca/video/10390476/demolition-begins-for-etobicokes-six-points-interchange-improvements.

26 Charles Komanoff, "Australia's Brief, Shining Carbon Tax," Carbon Tax Center, January 7, 2020, https://www.carbontax.org/blog/2020/01/07/australias-brief-shining-carbon-tax/.

27 Werner Antweiler, "Carbon pricing is essential," *University of British Columbia Magazine*, November 8, 2021, https://magazine.alumni.ubc.ca/2021/collective-wisdom/environment-opinion/how-get-net-zero.

28 After Karl Marx, who famously observed more than a century ago that history repeats itself, first as tragedy, then as farce.

Acknowledgements

THIS BOOK CAME ABOUT FOR A SIMPLE REASON. AS EXECUTIVE DIRECTOR OF THE Pacific Institute for Climate Solutions at the University of Victoria from 2009 to 2015, I worked closely with scholars, students, the business community, government ministers, legislators and a broad swath of the Canadian public, all focused on confronting the climate challenge so obviously affecting the long-term viability of society.

It was an exciting time. British Columbia's carbon tax had garnered international attention. Technological developments in the renewable energy arena were accelerating at a remarkable pace, everywhere. Canada had embarked on a massive expansion of its Athabasca oil sands, an initiative wrapped in a climate-changing downside that at the same time was bringing immense new wealth into the nation. Central to this ferment was the question of carbon emissions: how to curb them without incurring unacceptable economic damage; how to encourage businesses and families to conserve energy or produce it in different ways without harming their enterprises or lifestyles; how to make energy-production options more environmentally friendly while maintaining economic viability.

In reflecting on such questions I realized that an important story had yet to be told: *how* and *why* British Columbia—a province, not a nation—created a model for carbon pricing that was a template for the world, one that offered to put us all on the right road. That story demanded telling.

I am immensely grateful to so many who helped make this happen. Foremost is my encouraging and incredibly patient family. For more years than I care to mention, my wonderful wife, Carolyn Rymes, offered wise counsel, constructive insights and suggestions for grammatical, vocabulary and structural improvements that enhanced the prose. She is also highly attuned to the pulse of mainstream society, understanding apprehension generated by conflicting political messaging. Being able to tap into her sensitivity has been a great plus for me. Our computer scientist son, David—a whiz at proofreading—edited chapter drafts, banished excess commas, zeroed in on misspellings, and inserted semicolons when my sentences ran on. He frequently bailed me out when I didn't get along with my software. I salute both his attention to detail and his computer prowess. A dear family friend, Marge Berry, played an instrumental role when she asked me in late 2022 about the book's progress. "I haven't written a word for months," I said. "Well, get on it with it," she urged pointedly, of course in the nicest Canadian way. Her spur opened the gate; I sat down the next day and began pecking at the keyboard again. Thank you, Marge.

Michael Small, Canada's former high commissioner to Australia, generously opened governmental doors for me when I went to Australia in 2017 to learn how and why carbon-pricing efforts in that country had failed so badly. Thanks to him, I was able to interview a suite of prominent politicians and senior bureaucrats, some of whom expressly requested to remain unnamed, but to whom I offer my gratitude. During those few months Down Under, professors Brendan Gleeson and John Wiseman graciously accommodated me at the Melbourne Sustainable Society Institute of the University of Melbourne. My officemate there, political scientist Dr. Ben Parr, now at the University of Queensland, was instrumental in recommending books, papers and visual media sources that explored Australian political history. He was also a magical networker, advising me who in the Australian academic world I should interview. He greased the skids, as we say in Canada, and I remain very much in his debt.

I am grateful to all those in Australian political and academic circles who were willing to be interrogated by a Canadian interloper. That list is headlined by professors Frank Jotzo and the late Will Steffen (Australian National University, Canberra), professors Ross Garnaut and Don Henry (University of Melbourne), former Prime Minister Kevin Rudd (now Australian ambassador to the United States), graduate students Anita Talberg and Cathy Alexander (University of Melbourne), former Environment Minister Greg Combet, and Dr. Gordon de Brouwer, then secretary of the department of the environment and energy. All gave freely of their time. Noted environmentalist and professor Dr. Tim Flannery was exceptionally generous in inviting me to join his family at a spirited dinner at his Melbourne home one Sunday evening. Tim graciously committed to being interviewed while simultaneously helping to serve the meal and herd a bevy of youngsters who dashed in and out of the dining room.

Back in Canada, I benefited from conversations and interviews with many who were implicitly involved on BC's carbon tax during the heady years of the late 2000s: SFU and UBC economics professors Nancy Olewiler and David Green, respectively; Vancouver technology entrepreneur Jonathan Rhone; Martyn Brown, former chief of staff to the premier of BC; Clean Energy Canada CEO Merran Smith; Tzeporah Berman of Stand.earth; former BC finance officials Andy Robinson and Glen Armstrong; former cabinet ministers Mary Polak, Carole Taylor and Terry Lake; former premiers Gordon Campbell and Christy Clark; former Opposition Leader Carole James; former Opposition environment critic Spencer Chandra Herbert; Bill Tieleman of West Star Communications; and Linda Delli Santi of the BC Greenhouse Growers' Association. All agreed to discuss their perspectives and histories openly with me, as did some members of the BC Legislature who requested anonymity. All were candid, and I remain very appreciative. Others contributed indirectly, through provision of papers, data or discussions, in particular professors Kathy Harrison at UBC, Mark Jaccard at SFU and Stewart Elgie and Nic Rivers at the University of Ottawa.

Natural Resources Canada scientist Dr. Kathy Bleiker; University of Northern British Columbia forest entomologist Dr. Staffan Lindgren; UBC forestry professor Allan Carroll; Ken White and Tim Ebata of what was then known as the BC Forest Service; and Larry Pedersen, former chief forester of British Columbia, all contributed insights, observations, anecdotes and comments or illustrations that improved the early chapter on the mountain pine beetle. Thank you.

Harbour Publishing took a chance on this book, as tax policy isn't necessarily the most riveting of subjects. The team at Harbour was instrumental in getting my manuscript efficiently out the door. I am grateful to Ariel Brewster, Anna Comfort O'Keeffe, Luke Inglis, Zoë Mackenzie, Libris Simas Ferraz, Corina Eberle, François Trahan, Brian Lynch, Annie Boyar and John Montgomery for their many and varied contributions: coordinating, advising, editing, indexing, proofing, designing and marketing. Thank you all. I particularly salute copy editor and Toronto writer Michael Barclay. His excellent wordsmithing, intuitive capacity for phrasing, and attention to detail immeasurably improved the flow of my prose. Thank you, Michael.

Carbon taxation isn't always the most attractive topic of discussion with friends, relatives and dinner guests, but—surprise!—it seems to have come up repeatedly over the past several years, often leading to lively conversations. I have learned from them. Of the many who debated or discussed with me, I single out George Walton, Dorian Kachur, Lee Porteous, Virginia Rouslin, Shelley Forrester, George Abbott, Rollie Woods, my brother Dave and my lifelong friend Bryan Hardman for offering their perspectives. They helped me shape the arguments presented here.

And finally, my late father-in-law at Carleton University, economist and distinguished research professor T.K. Rymes, years ago stimulated my interest in the importance of pricing externalities. He reminded me many times that if we are to practice environmental stewardship properly, we must put a price on pollution. That perspective—his perspective—runs richly through this book.

Index

Note: Page numbers in **bold** indicate a figure. The letter *n* following a page number indicates an endnote.

A

Abbott, Tony
   on carbon tax as option, 97, 112, 124
   climate policy reversals, 123–25, 126, 128–29, 130, 132, 206–7
   and CPRS, 106–7, 109
   denial of and inaction on climate change, 97
   Direct Action Plan and ERF, 126–27, 128–29
   Emissions Trading System as carbon tax, 116, 119, 123–24, 126, 142–43
   vs. Gillard, 112–13, 116, 118, 120
   as leader and politician, 98–99, 118, 120, 123, 124, 127, 133
Adkin, Laurie, 181
air pollution, as problem, 21
Alberta, **5**, 34, 167, 198, 199
Alexander, Cathy, 100, 104, 111
Allan, John, 34
Andrews, Gwen, 82–83
Antarctica, 71, 89
Antweiler, Werner, 209
Armstrong, Glen, 43, 44
Australia
   cap-and-trade program, 92
   carbon dioxide emissions, 207
   carbon pricing (*see* carbon pricing in Australia)
   carbon tax (*see* carbon tax of Australia)
   and climate change (*see* Australia and climate change)
   Direct Action Plan, 126–29
   double dissolution or "double D" option, 103–4
   drought (Millennium Drought), 88–89, 119
   economy over environment, 74–75, 78, 82–83
   election of 2007, 85, 87–88, 90–91, 108
   election of 2010, 112–14
   electricity prices, 118–19
   emissions reduction, 118–19
   environment as concern, 68–70, 73–74
   environmental consciousness, 66–67, 70
   forest industry, 69–71
   global warming, 67–68, 85, 89
   "Green Paper" (2008), 92
   greenhouse gas measures and inaction, 74, 80–81, 82, 83

misogyny in politics, 120
nuclear option, 83, 84
Australia and climate change
    action, xi, 72, 73, 74–76
    climate commission, 124–25
    Climate Council, 125–26
    denial, 96–98, 100, 130
    inaction, 70–71, 75–76, 77, 78–83, 89–90, 102, 123, 131–32
    policy reversals, 123–25, 126, 128–29, 130, 132, 206–7
    politics of, 87
"Australia Clause," 80
Australian Greenhouse Office (AGO), 80–81, 82
axe the tax (axing of carbon tax)
    in Australia, 131–32
    and election of 2009 (BC), 141–42, 145–46
    as failure in BC, 145–46
    fiscal consequences, 200–201
    of NDP in BC, 135, 136–37, 138–39, 140, 142, 143–45, 146–48
    and Poilievre, 196, 200–202
    from politicians, 195

B

Bader, Maureen, 45–46
Baird, Julia, 132
Bauman, Yoram, 55
BC Liberal Party
    and carbon tax, 25–26, 141–42
    climate change action, 22–25, 150
    election of 2009, 64
    election of 2013, 155–56
    platform and views in early 2000s, 18–19
    talk on carbon tax by Inglis, 187–89

BCLICAT (British Columbia Low Income Climate Action Tax credit), 41–42, 190–91
Beijing, pollution in, 21
Belausteguigoitia, Juan Carlos, 192–93
Bennett, Bill, 15
Berman, Tzeporah, 138, 143–44, 177
Bjelke-Petersen, Joh, 70–71
Bjerregaard, Ritt, 80
Blair, Bill, 202
Bond, Shirley, 191–92
Bowen, Chris, 122
British Columbia (BC)
    carbon tax (*see* carbon tax of BC)
    climate change action, xi, xii, 15, 22–23, 24–25, 26–27, 29–30, 87, 153
    climate change as central issue, 19
    election of 2009, 64, 141–42, 145–46, 150
    election of 2013, 155–56, 162, 164–65
    election of 2017, 150, 183, 184, 185
    environmental consciousness, 58, 67
    fuel or gasoline (*see* fuels in BC)
    GDP and carbon tax, **58, 59**
    HST affair (sales tax harmonization), 36, 147, 150–51
    lobbying by oil and gas industry, 182
    mountain pine beetle (*see* mountain pine beetle)
    pipelines for oil and gas from Alberta, 149, 155, 156–58, 183–84
    politics and climate, 87
    regional planning, 15–16
    sustainability legislation, 21
    Throne Speech (2007), 22–23, 24–25, 26, 28–29, 30
    Throne Speech (2008), 29, 30
    winter temperatures, 8, 10

British Columbia Low Income Climate Action Tax credit (BCLICAT), 41–42, 190–91
British Gas, 63–64
Broecker, Wallace, x, 67
Brown, Bob, 69–70, 99–100, 101–2, 114, 115, 183
Brown, Gordon, 37
Brown, Jerry, 208
Brown, Martyn, 18–19, 22–23, 25–26, 30–31, 35, 39, 45
Bruce, Ian, 48
Burgess, Steve, 151
Bush, George W., 84
Business Council of BC, 48–49, 51

C

California, climate action and legislation on global warming, 22, 26, 84, 152, 208
Cameron, Ken, 16
Cameron, Max, 181
Campagnolo, Iona, 22
Campbell, Gordon
    action on atmospheric change, 17–18
    background, 15
    cap-and-trade program for BC, 27–28
    on C. Clark's 2017 loss, 185
    climate change action, 15, 21–23, 26, 29–30, 145
    on Dion's Green Shift, 34
    economy and environment, 18–19, 20
    election of 2009, 64, 145, 146, 150
    environment as concern, 15–16, 17, 18, 20, 72
    influential books for, 19–21
    on mountain pine beetle, 19
    as politician and mayor, 15, 17–18, 20, 35–36, 64, 151
    and regional planning, 15–16
    and sales tax harmonization (HST) affair, 36, 147, 150–51
    sustainability legislation, 21
Campbell, Gordon, and carbon tax in BC
    carbon tax as law, 32–33
    on gas prices after carbon tax, 56–57
    philosophy on carbon tax, 25, 33, 44–45
    planning of, 28, 29, 137, 138
    rate of carbon tax, 39
    reasons for success of tax, 108–10
    revenue neutrality, 29, 34–35, 40, 45
    work on tax, 32–33, 35
Canada
    climate change as concern in 1980s, 72, 73
    federal fuel charge, 197, 198–99
    refund of carbon tax (*see* refund of carbon tax)
    views on carbon tax from politicians, 196, 204–6
    See also carbon pricing and carbon tax in Canada
"Canada Carbon Rebate" label, 200
Canadian Association of Petroleum Producers (CAPP), 180–81
Canadian Taxpayers Federation, 45–46, 139–40
cap-and-trade programs, 27–28, 92–93, 94, 178–79
carbon dioxide emissions
    in Australia, 207
    cap-and-trade program of WCI, 27
    credits and limits in Canada, 197–98
    increase through LNG, 160, 170, 172
    reduction as goal in BC, 17, 160–61
carbon market, description of, 27

Carbon Pollution Reduction Scheme
    (CPRS, 2008)
  as choice, 93
  death of, 107–9
  description, 92, 93–94, 105
  introduction and attempted passing,
    93, 95–96, 98, 99–103
  problems with, 94–95, 98, 103–5
  reintroduction, 103–4, 105–7
carbon pricing and carbon tax in
    Canada
  "Canada Carbon Rebate" label, 200
  communication problems about,
    190–91, 199–200, 201–3
  exemption, 202–3
  federal politicians' views, 196, 204–6
  how it works, 197–98, 199
  introduction of, 197
  rate increase, 204–5
  refund to taxpayers, 190–91, 198–201
  revenue neutrality, 197–98, 200
carbon pricing in Australia
  failure of, 108–9, 123
  introduction, 84–85
  passing and repeal of, 117, 124, 126
  as yo-yo, 87, 110, 112, 127
  See also carbon tax of Australia
carbon pricing or trading (generally)
  importance, 209–10
  introduction, 84
  and revenue neutrality as concept,
    186, 187, 209–10
Carbon Tax Act (BC, 2008), 33, 60, 134–
    35, 238n341
carbon tax (generally)
  axing of (*see* axe the tax)
  communications about refund, 190–
    92, 193–94, 199–200
  conservative view, 33, 188–89
  description as simple scheme, 28
  design and implementation, 193
  seen as problem, 206–7, 208, 209
  selling approaches, 186–94
  talk to conservatives by Inglis,
    187–89
  as toxic phrase, 112, 116, 124, 142, 193
  and transportation, 189
  See also carbon pricing and carbon
    tax in Canada; carbon tax of BC
carbon tax of Australia
  overview as failure, xi, 64–65
  axing of, 131–32
  and J. Gillard, 113, 114–15, 116, 122, 206
  K. Rudd and CPRS, 93, 94, 95–96, 101,
    103–5, 106–9
  See also Abbott, Tony; Carbon
    Pollution Reduction Scheme;
    carbon pricing in Australia;
    Emissions Reduction Fund
    (ERF); emissions trading scheme;
    Emissions Trading System of
    Australia
carbon tax of BC
  anti-tax sentiment, 139–41
  carbon tax as phrase in Throne
    speeches and politics, 25, 28, 30–31
  climate-action dividend, 42
  communication problem with,
    190–92
  complaints and resistance, 45–48
  and consumption of petroleum
    products, 49–**57, 50, 55**
  economic impact, 49, **58**, 59, 62,
    63–64, 153
  and election of 2009, 64, 141–42,
    145–46
  and energy-intensive sectors, 59–62
  and environment, 64
  as example or template, 37, 192–94
  exemptions in, 38–39, 59–60

and filling up gas in the US, 54–55
foreshadowing in Throne Speech,
　25–26, 28–29
and G. Campbell (*see* Campbell,
　Gordon, and carbon tax in BC)
and GDP, **59**
how it works, 31, 38–39, 44–45, 163,
　207–8
and income tax, 35, 41, 46, 47, 137, 140,
　191–92
introduction, 31–32, 38
misinterpretation, 140, 145
in natural gas and LNG, 177–78, 179,
　180–81
as plan for BC, 17, 28
praise for, 46, 64
and price of gas, 55–58, 135–36, 138–39,
　141
and private sector, 44–45, 48
and prosocial behaviour, 54
rate, 39, **40**
rate increase, 39–40, 46–47
rate increase, freezing of, 50, 155, 162,
　163–64, 208
rate increase recommended by CLT,
　174
rate increase, reinstatement of,
　184–85
reasons for success, 108–10
refund through GST cheque, 190–91
reluctance of politicians, 23–24
revenue neutrality (*see* revenue
　neutrality in carbon tax of BC)
as success, xi, 64
as tax shift, 33, 38, 44–45, 47–48, 192
work on by government, 32–33,
　42–44
　See also Clark, Christy; James, Carole
Carr, Kim, 106
Cascadia states, climate action in, 208–9

cement industry in BC, 60–62
CFCs (chlorofluorocarbons), 71
Chandra Herbert, Spencer, 146, 148
Chubb, Philip, 90, 106, 107, 108, 113, 116,
　120
Chynoweth, Robert, 68–69
Clark, Christy
　carbon pricing for LNG, 174, 175,
　　176–79, 180
　carbon tax for natural gas, 177–78,
　　179, 180–81
　climate change action as plan, 171,
　　176
　climate change inaction, 75, 154, 176–
　　77, 183, 185
　consultation with oil and gas
　　industry, 180–82
　election of 2013, 155–56, 162, 164–65
　election of 2017, 150, 183, 184, 185
　environmental impact of natural gas,
　　170–71
　freezing of carbon tax, 50, 155, 162,
　　163–64
　hydrocarbon-based strategy, 157–62,
　　165–70, 171, 172, 179–80, 182–83, 208
　hydrocarbon-based strategy in
　　election, 149–50, 156–57, 165–66
　LNG as tool against global warming,
　　83–84, 161, 163, 182
　as politician, 153–56, 159, 161, 164–65,
　　183
　price of carbon tax, 50
　on revenue neutrality in carbon tax,
　　164
Clean Energy Bill (Australia, 2011)
　Emissions Trading System in, 117, 118
　as legislation, 117
　repeal of, 123, 126, 128, 129–30, 206
cleantech sector in BC, 62–63, 162–63

243

climate change
   action on, 14–15, 152–53 (see also individual people and places)
   as central issue, 19, 72–73
   cost of damage, 196
   impact, ix, xi, 13, 151–52
   warnings about, ix–x
   See also global warming
Climate Leadership Plan (CLP), 176–78, 179, 180, 182, 183
Climate Leadership Team (CLT) of BC, 171–76, 177, 178, 180, 181, 183
   members, 234n313
Clouds of Change report (1990s), 17
Coleman, Rich, 188
Collyer, Dave, 166–67
Colombia, carbon tax, 193–94, 237n334
Combet, Greg, 98–99, 104, 114–15, 117–19, 120
conservatives
   positive view of carbon tax, 33, 188–89
   spin on carbon tax in Canada, 196, 204–6
Coorey, Philip, 98
corporate income tax in BC, 41
Cosmos, Amor de, 86
Courchene, Thomas, 34
Coyne, Andrew, 205, 206
CPRS. *See* Carbon Pollution Reduction Scheme
Creating Our Future report (1990), 16

D
Daub, Shannon, 180
de Jong, Mike, 167–68, 170
Delli Santi, Linda, 60
Dendroctonus ponderosae. See mountain pine beetle
Denim Pine, 10–11

diluted bitumen ("dilbit"), 156, 157
Dion, Stéphane, 33–34
Direct Action Plan (Australia), 126–29

E
economy
   vs. environment as issue, 18–19, 20, 25–26, 74–75, 77–78, 82–83
   impact of carbon tax of BC, 49, 57, **58, 59**, 62, 63–64, 153
Ekins, Paul, 37–38, 64
Eltham, Ben, 128
Emissions Reduction Fund (ERF), as plan, 126–29
emissions trading scheme (Australia)
   and CPRS, 92
   failure of, 87
   as mechanism, 84–85
   as plan, 80, 81, 82, 83, 84, 91, 114–15
Emissions Trading System of Australia
   attacks and death of, 118, 119, 120, 122, 123–24, 126, 142–43
   draft and dealings, 115
   passing, 117
   questions about, 115–16
   success and benefits, 118–19, 123
Emissions Trading System of EU, 27, 92–93, 115
energy-intensive sectors, and carbon tax of BC, 59–62
environment
   vs. economy as issue, 18–19, 20, 25–26, 74–75, 77–78, 82–83
   public policy, 16–17, 18, 73–74
European Union, 27, 84, 92–93, 115
Evans, Chris, 100
Ewart, Heather, 116

F
"federal fuel charge," 197, 198–99

Finlayson, Jock, 48–49, 51, 54
Flannery, Tim, 20, 124–26
Fleming, Rob, 146
Fletcher, Robson, 199
forest fires suppression, 11
forest industry
    in Australia, 69–71
    in BC, 10–11, 12–13
    impact on fossil fuel consumption, 51
fossil fuels
    "federal fuel charge," 197, 198–99
    impact of carbon tax on consumption, **49**, 49–55, **53**, **56**, 58–59
    See also fuels in BC; oil and gas
Fourier, Jean-Baptiste Joseph, x
fracking in BC, 158, 159
Frankel, Alex, 124
Friedman, Milton, 15
fuel in Canada, "federal fuel charge," 197, 198–99
fuels in BC
    change in consumption, 49–60, **50**, **55**, **58**, **59**
    change in sales, **58**
    cost with carbon tax, 39, **40**
    filling up in the US, 54–55
    price after carbon tax, 55–58, 136
    price rises before carbon tax, 135–36, 138–39
    price rises due to market, 58
    sales in BC before and after carbon tax, **58**
    world prices, 55–56, 141
furnace oil, 202–4

G

Garnaut, Ross, 68, 89, 91, 93, 114
Garrett, Peter, 67
gas or gasoline in BC. *See* fuels in BC
General Assembly of the United Nations, 153
Gibson, Gordon, 17
Gillard, Julia
    vs. Abbott, 112–13, 116, 118, 120
    and carbon tax or pricing, 113, 114–15, 116, 122, 206
    climate pricing, 112
    and CPRS, 95, 103
    and Emissions Trading System, 115–17, 118
    Kyoto Protocol ratification, 91
    as leader/PM and in election, 108, 111–14, 119–21
    misogyny against, 120
glaciers in BC, and climate change, ix
global recession of 2008–09, 50–51
global warming
    in Australia, 67–68, 85, 89
    foreshadowing of, ix–x
    impact on mountain pine beetle, 8, 10, 13
    legislation in California, 22, 84, 208
    LNG as tool against, 83–84, 161, 163, 182
    as phrase and concern, 67
    as reality, x–xi, 151–52, 196
    See also climate change
goods and services tax (GST), and refund cheque, 190–91
Gordon, Julie, 169
Gore, Al, 131
Grattan, Michelle, 90
Great Barrier Reef, 77
Greater Vancouver Regional District (GVRD), 15–16
Green, David, 137–38
Green Fiscal Commission (UK), 37–38

Green Party (Australia)
    and CPRS, 95, 99–100, 101–3, 106
    and Emissions Trading System, 114–15, 116–17
    seats in government, 90, 102, 113–14
Green Party (BC), seats in government, 184
Green Shift program (of S. Dion), 33–34
greenhouse gas emissions
    action on in BC, 17, 24–25
    impact, ix–x
    increase as problem, 16–17
    measures and inaction in Australia, 74, 80–81, 82, 83
    reduction targets, 24–25, 73
Greenhouse Gas Pollution Pricing Act (Canada, 2018), 197
greenhouse vegetable industry in BC, 59–60
GST (goods and services tax), and refund cheque, 190–91

H

Hamilton, Clive, 76, 78, 80
Harcourt, Mike, 16
Hardwick, Walter, 16
Harper, Stephen, 34
Harrison, Kathryn, 145–46
Hawke, Bob, 68, 70–71, 73–74, 75–76
heating oil, 202–4
Hébert, Chantal, 206
Helliwell, David, 63
Henry, Don, 94, 104, 107, 131
Hill, Robert, 78, 79–81
Horgan, John, and government, 140, 184, 238n341
Howard, John
    and Australian Greenhouse Office (AGO), 80
    emissions trading scheme, 84–85
    and global warming, 85
    inaction on and disinterest in climate change, 76–77, 78–79, 80–83, 88, 89–90
    newspaper columns, 76
    nuclear option, 83, 84
    political career, 76, 85, 87–88
Huhne, Chris, 152
Hunt, Greg, 124, 125

I

income tax
    in Australia, 117–18
    and carbon tax of BC, 35, 41, 46, 47, 137, 140, 191–92
    reduction through carbon tax, 193–94
    See also tax
infrared radiation, impact of, x
Inglis, Bob, 186–89, 193
Inslee, Jay, 208

J

Jaccard, Mark, 177
Jackson, Peter, 3
James, Carole
    axe the tax campaign, 135–37, 138, 140, 142, 143–45, 146–47, 148
    as leader of NDP, 147–48
    support of carbon taxes or pricing, 135
    vote against Carbon Tax Act, 134
Jeffery, Michael, 91
Jensen, Dennis, 105
Johnson, Jim, 52
Johnson, Lester, 1
Jones, Tony, 130
Jorgensen, Peter, 80
Jotzo, Frank, 128
Joyce, Barnaby, 97

K

Keane, David, 180
Keating, Paul, 73, 74, 75, 76
Kelly, Paul, 109
Kelly, Ros, 75, 83
Kinder Morgan, 183–84
Kines, Lindsay, 138
Kirk, Judy, 16
Kitzhaber, John, 208
Komanoff, Charles, 207
Krugman, Paul, 191
Kurl, Shachi, 200
Kusmierczyk, Irek, 205–6
Kwan, Jenny, 148
Kyoto Conference (1997), 78–80
Kyoto Protocol, ratification by Australia, 82, 83, 90, 91

L

Laffer, Arthur, 187
Lake, Terry, 157–58, 161, 163, 171, 183, 187, 188
Latin American and Caribbean Carbon Forum (2014), 192–94
Lawley, Chad, 55
Lawson, Jamie, 144
Les, John, 188
Lewis, Stephen, 72
Liberal Party (Australia), denial of and inaction on climate change, 77, 78, 96–97, 98
Liberal Party (BC). *See* BC Liberal Party
Lindgren, Staffan, 3
Littlemore, Richard, 134, 135, 147, 148
LNG (liquefied natural gas). *See* natural gas and LNG
lumber in BC, 10–11, 12–13

M

McDonald, Jessica, 30
Macfarlane, Ian, 96, 98, 99, 101
McKenna, Catherine, 197
McKenzie, Amanda, 125
MacKinnon, Janice, 34
Macklem, Tiff, 202
McMartin, Will, 147–48
MacNab, Josha, 177
McSweeney, Michael, 62
M'Gonigle, Michael, 21
Marr, David, 99
Mason, Gary, 144
May, Elizabeth, 71–72
Megalogenis, George, 76
Menounos, Brian, ix
methane, 158, 166
  See also natural gas and LNG
Millennium Drought in Australia, 88–89, 119
Milne, Christine, 101, 102, 129
Minchin, Nick, 96–97, 98, 100
Moe, Scott, 205, 206
Monbiot, George, 20–21
Montreal Protocol, 71–72
mountain pine beetle (*Dendroctonus ponderosae*)
  area infested, 4, **5**
  and blue-stain fungus, 10
  boring method and result, 6, 8
  as cause for action, 19
  damage to trees, 2–3, 4, **7**, 8, **9**, 11, 12
  description and life, 1, **2**, 4, 6, 8, 10
  expansion of epidemic, 11–12
  impact on mills and revenue, 12–13
  population explosion in BC, 1–4
  winter survival, 8, 10
Mulroney, Brian, 72
Murdoch, Rupert, and press of, 96, 106, 109, 115, 118, 132
Murray, Joyce, 18

N

natural gas and LNG
  and carbon dioxide emissions, 160–61, 170, 172
  and carbon pricing, 174, 175, 176–79, 180
  and carbon tax of BC, 177–78, 179, 180–81
  demand and prices, 158–59, 160
  environmental impact, 170–71
  LNG as tool against global warming, 83–84, 161, 163, 182
  LNG tax, 162, 167–70
  production and export in BC, 155, 159–62, 167
  reserves in BC, 149
  in strategy of C. Clark government, 158–62, 166, 167–70, 179–80, 208
  use in greenhouse industry, 60
NDP (or New Democrats) of BC
  axe the tax stance and campaign, 135, 136–37, 138–39, 140, 142, 143–45, 146–48
  against carbon tax, 45, 46
  and rural vote, 144, 145
  shock at their anti-tax stance, 137, 138, 139, 143
  vote against Carbon Tax Act, 134–35
neutralization of revenue. *See* revenue neutrality; revenue neutrality in carbon tax of BC
A New Era for British Columbia (platform of BC Liberal Party), 18
Northern Gateway Pipeline, 157
nuclear option, in Australia, 83, 84

O

oil and gas
  industry consultation with C. Clark government, 180–82
  pipeline expansion/twinning, 149, 155, 156–58, 183–84
  See also fossil fuels
Olewiler, Nancy, 137, 138, 177
Oregon, climate action, 208
O'Toole, Erin, 204
"Our Changing Atmosphere: Implications for Global Security" conference (1988), 16–17
Output-Based Pricing System (OBPS), 197–98
ozone hole and layer, 71, 89

P

Palmer, Clive, 129, 130–31
Palmer, Vaughn, 146, 150, 169
Palmer United Party (PUP), 129, 130, 131
passenger vehicles, impact of carbon tax of BC on, 52
Pearse, Guy, 77–78, 81, 82, 84
Pilon, Denis, 148
pine trees
  attacks by mountain pine beetle, 6, 8, 10, 12
  blue-stain fungus, 10–11
  resins against attacks, 6, **7**, 8
Planet U (M'Gonigle and Starke), 21
Poilievre, Pierre, 196, 200–202, 203, 204
Polak, Mary
  carbon tax for natural gas, 181
  and CLP, 176, 180
  and CLT, 173, 176, 177–78, 179–80
  on freezing of carbon tax, 163, 177, 178
  and revised tax regime, 169, 170, 178–79
  transportation and carbon tax, 189
  twinning of pipeline, 157
Pont, Lynn, 10

pricing of carbon. *See* carbon pricing and carbon tax in Canada; carbon pricing in Australia; carbon pricing or trading
private sector, and carbon tax of BC, 44–45, 48
provinces, and federal carbon tax plans, 34
pulp and paper industry, 51–52
Pulse Energy, 63

Q
Queensland, forest industry, 70–71

R
radiant heat, impact, x
Ralston, Bruce, 46
rapid transit, 51, 52
Raupach, Michael, 132
Reagan, Ronald, 23
Red Sky at Morning (Speth), 19–20
Rees, William, 21
refund of carbon tax
   communications about, 190–92, 193–94, 199–200
   consequences of axing the tax, 200–201
   as system in Canada, 190–91, 198–201
Reimer, Andrea, 139
Renewable Energy Target (RET), 131
revenue neutrality
   in BC (*see* revenue neutrality in carbon tax of BC)
   in carbon pricing and carbon tax in Canada, 197–98, 200
   in carbon pricing generally, 186, 187, 209–10
   and inflation, 202, 239n350

   and return to taxpayers, 198–99, 203, 204, 206
   seen as cash back, 188, 192
revenue neutrality in carbon tax of BC
   benefits, 45, 63, 164
   and freezing of carbon tax, 163–64
   how it works, 34–35, 40–41, 47–48
   and income tax, 35, 41, 46, 47, 140, 191–92
   in legislation, 31
   planning and design of, 28, 29, 32–33, 44
   revenue returned to taxpayers, 41–42, 198–99, 238n341
Rhone, Jonathan, 62–63, 162–63
Richardson, Graham, 69–71, 72, 73
Rivers, Nic, 53–54, 59
Robinson, Andy, 43–44
Rudd, Kevin
   climate action, 88, 89, 91–92, 103, 107–8, 122
   and CPRS, 93, 94, 95–96, 101, 103–5, 106–9
   and Emissions Trading System, 122
   and J. Gillard, 108, 112, 119, 120, 121–22
   Kyoto Protocol ratification, 91
   political career and election, 87–88, 90–91, 108, 109, 112, 121, 122, 123

S
Saskatchewan, 34
Schaufele, Brandon, 53–54, 59
Schlossberg, Jack, 189
Schreck, David, 136
Schwarzenegger, Arnold, 22, 23, 26, 84, 208
"Securing Our Energy Future" (Australia, 2004–05), 81, 83
Shanahan, Dennis, 120

Shergold, Peter, and Shergold report, 84–85
Simpson, Bob, 121, 148
Simpson, Jeffrey, 121
Skuce, Andy, 55
Smith, Danielle, 205–6
Smith, Merran, 171, 172, 183
Smith, William, 86
Smyth, Michael, 136, 164–65
South Carolina, global warming and politics, 186–87
Speth, James Gustave, 19–20
Starke, Justine, 21
"Statement on the Environment: Our Country, Our Future" (Australia, 1989), 73–74
Steffen, Will, 95, 114
Stern, Nicholas, 69
Summers, Anne, 120
Suzuki, David, 142
Swan, Wayne, 106

T

Tasmania, forest industry, 69–70
tax
    BCLICAT credit, 41–42, 190–91
    carbon tax as toxic phrase, 112, 116, 124, 142, 193
    GST refund cheque, 190–91
    tax as word in politics, 23–24, 30, 116, 124
    tax shift of carbon tax of BC, 33, 38, 44–45, 47–48, 192
    See also axe the tax; income tax
taxing of carbon. *See* carbon pricing and carbon tax in Canada; carbon tax (generally); carbon tax of Australia; carbon tax of BC
Taylor, Carole, and carbon tax of BC
    description of tax, 38
    introduction of, 31–32, 38
    and private sector, 48
    on rate of tax, 39
    revenue neutrality, 40–42, 44
    work on tax, 32–33, 43
Taylor, Maria, 73, 74–75
Thatcher, Margaret, 153
Thivierge, Vincent, 55
Thomson, Graham, 177
Thomson, Kelvin, 82
Tieleman, Bill, 139–41, 142, 148, 151
Toronto conference (1988), on climate change, 72–73
"the Toronto target," 73, 75
Trans-Mountain Pipeline, 149, 155, 156–58, 183–84
Trudeau, Justin, and government, 196–97, 199–200, 202–4
Turnbull, Malcolm, 90, 95–96, 98–99, 105, 128, 133
Tyndall, John, ix–x

U

ultraviolet radiation (UV), 71
United Kingdom, interest in carbon tax of BC, 37
United States, climate action in Cascadia states, 208–9
uranium, 83

V

Voters Taking Action on Climate Change (VTACC), 137

W

"war in the woods" at Clayoquot Sound, 67
Washington State, climate action, 208
Weaver, Andrew, 138, 142, 144, 165, 166, 179, 184

Western Climate Initiative (WCI), 26–27
White, Ken, 3–4
Whitmarsh, Graham, 30
wildfire suppression, 11
Wilkinson, Jonathan, 201
Wilkinson, Marian, 69, 70
Wilson, Gordon, 17
Wong, Penny, 96, 98, 99, 101, 102

Y
Yap, John, 37
Yunker, Zoe, 180

Photo by Jennifer Pedersen

THOMAS F. PEDERSEN, PHD, FRSC, FAGU, IS A PROFESSOR EMERITUS at the University of Victoria. He holds a PhD in marine geochemistry and is internationally recognized for his research in the history of the oceans and the response of the sea to climate change. He is the former Dean of Science at University of Victoria and served as the founding Executive Director of the Pacific Institute for Climate Solutions. Pedersen has made significant contributions to climate-change policy and analysis in British Columbia and Canada, served on multiple international committees and has received numerous honours for his scholarship. He has coedited two books on global change and published over 130 scientific papers and book chapters. Pedersen lives in Victoria, BC.